The Homeland of the Aryans

Evidence of Ṛigvedic Flora and Fauna & Archaeology

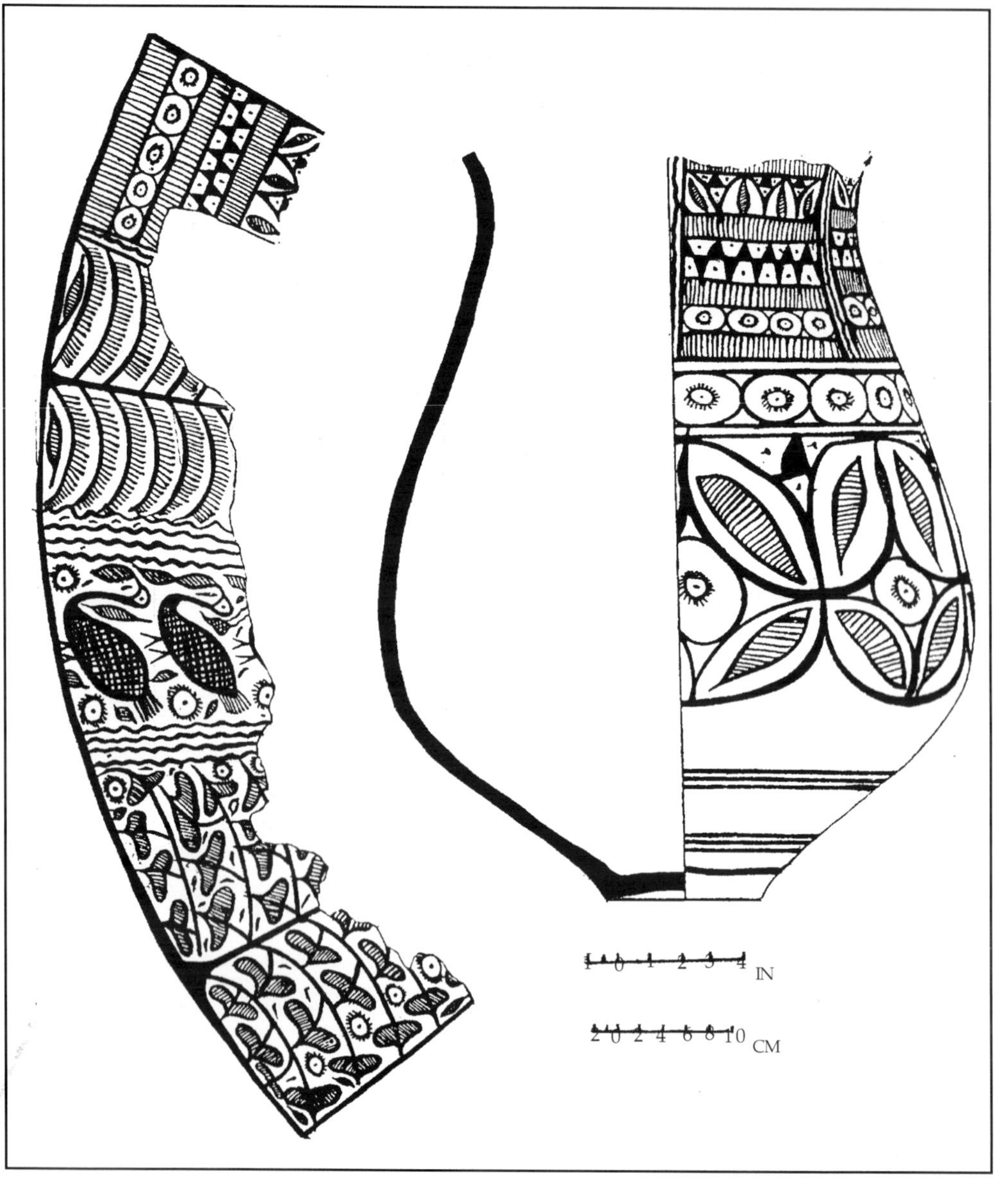

Lothal: Painting on a vase showing *Pīpala* leaf and peacock.
Mature Harappan.

The Homeland of the Aryans

Evidence of Ṛigvedic Flora and Fauna
& Archaeology

B.B. LAL

with contributions by
K.S. SARASWAT

Aryan Books International
New Delhi

THE HOMELAND OF THE ARYANS
Evidence of Ṛigvedic Flora and Fauna & Archaeology

ISBN: 978-81-7305-283-5

Published in 2005 by :
Aryan Books International
Pooja Apartments, 4B, Ansari Road, New Delhi-110 002 (India)
Tel.: 23287589, 23255799; Fax: 91-11-23270385
E-mail: aryanbooks@gmail.com
www.aryanbooks.com

Printed in India

Jacket illustrations: The *pīpala* and peacock are amongst the outstanding examples respectively of the flora and fauna of the *Ṛigveda*. These are also amongst the most favourite motifs on the Early/Mature Harappan pottery.

To
The Late Professor Walter A. Fairservis
and his distinguished student,
Professor Gregory L. Possehl,
to
The Late Professor George F. Dales
and his distinguished student,
Professor Jonathan Mark Kenoyer,
and to
Professors Jim G. Shaffer and J.-F. Jarrige,
all of whom – located far away,
have contributed so much
to the archaeology of the Indian Subcontinent,
is this little book dedicated

CONTENTS

PREFACE

I propose to begin this little book on a somewhat lighter note since what will follow in the subsequent pages is going to be tough and, maybe, even tiring. It was the winter of 1960 (I don't remember the year exactly, but it was thereabouts). I was still serving the Archaeological Survey of India and was at that point of time the Director of its School of Archaeology. A senior archaeologist from the United States of America, Prof. Walter A. Fairservis, was on a visit to India. We were good friends and I invited him to a dinner at my residence in New Delhi. He was accompanied by a young student, Gregory L. Possehl, whom I had never met before. I also requested this young man to come along for the dinner, an invitation he kindly accepted. We had typical Indian food – vegetarian and non-vegetarian curries, other vegetables, *rāitā*, salads, *pāpaḍam* and even some pickles. The main course was, as usual, followed by a couple of sweet dishes, one of which was specially made by my wife. She prepares discular, somewhat biconvex, cake-like pieces, from *maidā* (a wheat-product), fries them in *ghee* (clarified butter) and then soaks them in sugar-syrup. This preparation was immensely relished by the guests. After the dinner while we were talking over a cup of tea, a nasty idea entered my mind. Sometime earlier, during my visit to Jaipur, I had bought some dry fruits – cashew nuts, almonds (without the shell), etc., which, though made of wood, looked perfectly like real ones. I took these 'fruits' out of the shelf and placed them before the guests. Poor chaps – they were struggling hard to crack the nuts, but these wouldn't yield a bit. Having watched the fun for a while I greatly apologized to the guests for my mischief and the guests were gracious enough to forgive me.

Alas! my dear old friend Walter is no longer alive to share the memories of this fun, but the then young man, Greg, now a very distinguished Professor at the University of Pennsylvania, Philadelphia, and one of the few leading American scholars on the Indus Civilization, recalls this (somewhat mischievous) incident with nostalgia. In fact, on a couple of occasions he himself recounted it when a few of us got together. Greg, I am sure, would also forgive me for yet another uncharitable deed. He recently published an over-1000-page-thick book, *Indus Age – The Beginnings.* It is the most remarkable book of its kind – a storehouse of rare data with an in-depth analysis thereof. However, out of sheer humour and just to tease Greg a few common friends and I told him that the book was a befitting tribute to his 'calibre' – both intellectual as well as physical! It is no secret that fat persons have a great sense of humour and our friend too has that in ample measure. He thus took our joke about his book in the spirit it was meant. The book is indeed a monumental piece of work deserving all praise and congratulations.

The readers and surely Greg would be wondering as to why I have narrated this dinner-incident. In my Introduction to this book I am going to quote a paragraph from another book of Greg, namely *Indus Age – The Writing System,* which has in a way 'inspired' (not provoked) me to write the present book. Though my comments are not going to be very palatable, like the dry fruits already referred to earlier, yet I am confident that my dear friend Greg will not take them amiss. In fact, I would be only too glad to receive his counter-comments. It is only through scholarly dialogues that the truth will reveal itself. So be it.

I have also included in this book two other papers of mine, related to the problem of 'The Aryan Homeland'. One of the papers, captioned 'The Homeland of Indo-European Languages and Culture: Some Thoughts', was presented at a Seminar held on January 7-9, 2002 in New Delhi, under the auspices of the Indian Council of Historical Research. Since the publication of the proceedings of this Seminar is likely to take a long time and the paper is considered seminal to the 'homeland problem', it is being included here.

The other paper bears a somewhat unusual title, namely 'Should One Give Up All Ethics for Promoting One's Theory ?' In an utter desperation to establish that the Aryans migrated into India from elsewhere – Central Asia and the like, many scholars even go to extent of wantonly misinterpreting the Sanskrit texts. This paper exposes one such attempt, namely that by Prof. Michael Witzel of the Harvard University, USA. Though this paper too is scheduled for publication in the *EAST and WEST,* yet it is being included here since its publication in that journal is also likely to take considerable time. I trust the aforesaid papers, like the main essay of this book, will set the readers re-thinking. Maybe they do not agree with me straightaway, but let them ponder over the whole issue, including that of 'using unfair means' to establish a preconceived theory.

Los Angeles: **B.B. Lal**
Vijaya Daśamī
October 4, 2003

ACKNOWLEDGEMENTS

First and foremost, I would like to record my deep sense of gratitude to Dr. K.S. Saraswat, a Senior Scientist (Scientist F) at the Birbal Sahni Institute of Palaeobotany, Lucknow. Besides supplying me colour illustrations of most of the trees and plants included herein, he has done the arduous job of giving me comprehensive data about their provenance. Without this data it would have been impossible for me to argue the case.

I am obliged to the Indian Council of Historical Research for letting me publish here the paper, 'The Homeland of Indo-European Languages and Culture: Some Thoughts', which I presented at a seminar held under its auspices at New Delhi from 7-9 January, 2002. (The publication of the proceedings of the Seminar is likely to take some time.) As regards another paper of mine, viz. 'Should One Give Up All Ethics for Promoting One's Theory', which was sent quite some time ago for publication in the *EAST and WEST,* I am sorry to record that the editor raised some peripheral and odd points. Though my clarification was duly sent to him, I have not received any communication from him so far. Anyway, since the publication of the paper in that journal is also likely to take time and the matter is quite vital to the 'Aryan Homeland' debate, I am taking the liberty of including it in the present book, thanking in advance the editor of the *EAST and WEST.*

Shri J.P. Joshi, my friend and colleague in the Archaeological Survey, has been kind enough to provide me with the references from the *Śivamahāpurāṇa* and *Raghuvaṁśa* cited in Appendix 3. To Prof. Makkhan Lal, Director, Delhi

Institute of Heritage Research and Management, I am indebted for lending me many a book which I needed from time to time. Besides, he has been instrumental in procuring some of the illustrations. Thanks are also due to Shri D.V. Sharma, Superintending Archaeologist, Excavations Branch, ASI, New Delhi, for kindly lending the services of Shri Vinod Kumar Srivastava for preparing some of the photographs.

Also, Shri S.K. Sharma has helped me in the preparation of the line-drawings, in particular the map showing the overlap of the Vedic and Harappan terrains. Shri Vishnu Kant has lent his assistance in proof-reading and indexing. To both of them I am most thankful.

My friend, philosopher and guide, Shri Krishna Deva, is no longer amidst us. Though, I feel very lonely, nevertheless, his memory continues to inspire me. May his soul rest in peace!

Thanking my wife, Kusum, is something which she does not like. Yet I cannot help doing that, since I have found her offering valuable comments which have often guarded me from going off the rails. I would also like to place on record the fact that my eldest son, Air Vice Marshal Rajesh Lal, has helped me in proof-reading, besides making some important suggestions.

And finally the enthusiastic publisher, Shri Vikas Arya, who did a very good job of my last book, *The Sarasvatī Flows On: The Continuity of Indian Culture.* I am sure he will spare no efforts to make this one equally presentable.

LIST OF ILLUSTRATIONS

CHAPTER I

INTRODUCTION

For more than a century now it is being orchestrated that the Aryans* came to India from outside, more specifically from Central Asia. In my book, *The Sarasvatī Flows On,* I thoroughly examined the issue and showed that there was no basis for such an assumption; and no one has yet come forward countering my arguments. However, there is still another ground on which the 'Aryan Invasion/Migration' theory is being bolstered up. It relates to the names of trees in the Indo-European languages. It has been stated that these names include certain species which pertain to cold climate and since India is a country of hot climate these trees must take us to a cold-climate area in search of the 'original home'. For example, Possehl (1996: 65) holds as follows:

> There has been a great deal written about Aryans and Indo-Europeans, much of which is either confused or confusing. Even recent books on these peoples, while they receive much attention, can be old fashioned (e.g. Renfrew 1987). One thing seems certain; the speakers of Vedic Sanskrit are the earliest *well documented* [italics in the original] speakers of an Indo-European language in the subcontinent, and they came from elsewhere. This conclusion comes from a number of sources, the two most important of which are: (1) their books, which tell us that they were in new lands filled with non-Aryan peoples and (2) Indo-European words for trees

---

* For a very long time this term has been taken to signify a 'race', but now it is being progressively realized that it is a misconception.

which are species such as birch, Scotch pine, linden, alder and oak. These are plants from a temperate environment and the fact that their names are shared among the early languages of the family suggests a homeland in this environment (Friedrich 1970, especially pages 152-58).

I do not propose to deal right here *in extenso* with Possehl's first assertion, viz. 'their books [by which he evidently means the books of the Indo-Aryans], which tell us that they were in new lands filled with non-Aryan peoples'. In this particular context, the readers' attention is invited to Chapters V and VI (pp. 63-88) that have been specially appended to present my viewpoint. However, it is Possehl's second statement about the cold-climate trees that is going to be the mainstay of this book. Just to put in a nutshell what is going to be given in detail later on, there is ample evidence to show that: **(1) The earliest known book of the Aryans, viz. the *Ṛigveda*, does *not* mention any of the species of cold-climate trees so confidently enumerated by Possehl, and (2) on the other hand all the species of trees mentioned in this earliest text of the Aryans belong *not to a cold climate but to a tropical one.* As would be clear from the pages that follow, the provenance of these trees does not go west of Afghanistan and is by and large confined to what is now India, Pakistan and Bangladesh, with a spill over to Sri Lanka, Myanmar and a bit further to the east. These data completely knock the bottom out of the 'cold-climate thesis' for the 'original home' of the Aryans.**

Also included in this book is a discussion on the fauna of the *Ṛigveda*. The occurrence in this earliest text of the Aryans of names of such species as the lion, camel, elephant and peacock, which essentially belong to a tropical and sub-tropical climate, and the total absence of any species specifically of the cold regions from the Ṛigvedic list clearly show, once again, that the 'cold-climate-home' thesis for the Vedic Aryans is baseless. Indeed, as I have stated elsewhere (Lal 1997: 281-87), **'It is time to Rethink'**.

CHAPTER II

THE FLORA

For the sake of easy comprehension, the flora is being dealt with here under two separate sub-headings, viz. (A) Trees and (B) Plants and Grasses. Under each sub-heading, the individual items will be taken up in the order of their initial alphabet in the Devanāgarī script. The Roman-alphabet serial could have as well been applied but that was not possible since in many cases the Latin name is still unclear or remains unidentified. Anyway, nothing is lost by adopting the serial order of the Devanāgarī script.

It needs also to be added that not each and every tree/plant/grass that finds a mention in the *Ṛigveda* is being included here. The reason is twofold. In the first place, it has not been possible to identify with certitude many of the trees, e.g. *kākambīra* (*RV* 6.48.17), and the problem is much more complicated in the case of plants and grasses, in particular the latter. The position is still worse in the case of medicinal plants, like *aśvāvatī, somāvatī*, etc., mentioned in *RV* 10.97.7. Then there are certain items about the identity of which there is even greater confusion: for example, about *araṭu* (*RV* 8.46.27). Not only is it debated whether it was a tree or a plant, but there are also some scholars who see in the word a proper name. The second reason is that for the purpose of the present discussion, not all the trees/plants/grasses are of importance, since many of them are of such common occurrence all over the world that going into their detailed distribution-pattern would be of little use in the present context. Nevertheless, care will be taken to include in this discussion, as far as possible, all the significant items.

A. TREES

(i) *Aśvattha (Ficus religiosa* L.)

Also known as *pīpala* in Hindi, it is a large-sized tree, rising to a height of over 30 m (Fig. 2.1). Its wood seems to have been used for making vessels which would hold the Soma (***RV* 1.135.8**). In later texts it is stated that of the two sticks (*araṇī*) of wood that were used to kindle fire for sacrificial purposes, the upper one was that of the *aśvattha*, since its wood was harder than that of the lower one which was that of *śamī*. In a hymn dedicated to the praise of herbs (***RV* 10.97**), it is mentioned that the *aśvattha* tree as well as the *parṇa* (*Butea monosperma*: Syn. *Butea Frondosa*) were the homes of the herbs: it may mean that the juice of the herbs was kept in vessels made from the wood of these trees, to be used in the sacrificial rituals. The relevant passages are quoted below:

***RV* 1.135.8:**

atrāha tad vahethe madhva āhutim
*yam**aśvatthamu**patiṣṭhanta jāyavo-asme te santu jāyavaḥ* /
Sākam gāvaḥ suvate pachyate yavo
na te vāya upa dasyanti dhenavo nāpa dasyanti dhenavaḥ //

Ride hither to the offering of the pleasant juice, **the holy Fig-tree** which victorious priests surround: victorious be they still for us.
At once the cows yield milk, the barley-meal is dressed. For thee,
O Vāyu, never shall the cows grow thin, never for thee shall they be dry.[1]

***RV* 10.97.5:**

Aśvatthe *vo niṣadanam* ***parṇe*** *vo vasatiṣkṛitā* /
Gobhāja it kilāsatha yat sanavatha pūruṣam //

[O herbs], the **holy Fig (Aśvattha) tree** is your home, your mansion is the **parṇa** tree:

---

1. The English translation has been taken all through from Griffith, Reprint 1995, though sometimes it has been found to be faulty, as pointed out at the respective places.

Fig. 2.1 *Aśvattha* (*Pīpala, Ficus religiosa* L.)

Winners of cattle ye shall be if
ye regain for me this man.

An interesting feature in the case of the *pīpala* tree deserves mention. Birds eat its fruits and the hard seeds pass out undigested in the excreta. When the excreta is dropped in the cracks of the bark of other trees the seeds germinate and produce a *pīpala* plant thereon. This feature of its growth on other trees, seems to have been noted way back in antiquity, by the authors of the *Atharvaveda* (cf. *Av*, iii. 6).

This tree is held sacred by Buddhists and Hindus both. Lord Buddha is said to have received enlightenment while meditating under an *aśvattha/pīpala/bodhi* tree at Bodh Gaya in Bihar. In the third century BC, a sapling of this Bodh Gaya tree was taken to Sri Lanka and planted at Anuradhapura. It is held that the present-day Bodhi tree over there is the same as the one planted some 2200 years ago. Whether this is so or not is hardly relevant, since the fact remains that the tree is held sacred because of its association with Lord Buddha. In India, the Hindus also worship this tree. However, it is interesting to find that at the cremation-ground there is usually a *pīpala* tree to which is hanged an earthen pot, filled with water – a ritual associated with the dead.

As regards the geographical distribution of the *aśvattha*, Saraswat has communicated as follows:

> Wild in the sub-Himalayan forests from the Punjab eastwards; Bengal, Orissa, central India; planted and runs wild in other parts of India; Burma (Myanmar); Sri Lanka (Gamble 1902; Brandis 1906). Planted or self-sown in the plains of Punjab, Sindh and Lower Baluchistan, sometimes ascending to a height of about 1200 metres. Also occurs in Kashmir foothills (Stewart 1972). Cultivated in Jalalabad region of Afghanistan (Kitamura 1960). Rarely, the tree also occurs in the Malayan region (Hooker 1878).

Archaeologically, we know that, whether held sacred or not, this tree drew attention of ancient Indians since the fourth millennium BC, if not earlier. Its leaves (cf. Fig. 2.2) have been painted on the Early Harappan (ca. 3000-2800 BC) pottery from Kunal in District Hissar, Haryana. The painting is bichrome. Thus, while the outline of the leaf and some parts in the interior are in black colour, a major portion of the inner space is filled with white (Fig. 2.3). This kind of use of two colours was a distinctive feature of the Early Harappan pottery – from Kunal in the east to Kot Diji in Sindh in the south-west, to Rehman Dheri in the North-West Frontier Province of Pakistan. By the Mature Harappan times (ca. 2600-2000 BC), the *pīpala*-leaf motif, along with a few others, became the hallmark of the painted pottery (Fig. 2.4). Further, charcoal pieces of *pīpala*-wood have been found in the Early Harappan levels of Kunal (ca. 3000-2800 BC) and in the Mature Harappan deposits at Banawali and Rohira. This clearly affirms the presence of the *pīpala* tree in Haryana and Punjab since the ancient times.

We have stated earlier that at present this tree occurs as far west as Afghanistan. In antiquity, the case was no different. From the site of Mundigak come stemmed cups, painted on the exterior with *pīpala*-leaves (Fig. 2.5 here; and Casal 1961). These are datable back to the middle of the third millennium BC.

(ii) *Kiṁśuka / Parṇa* (*Butea monosperma [*Lamk.*]* Taub., Syn. *Butea frondosa* Roxb. / *Butea superba* Roxb.)

It is a moderate-sized tree bearing red-coloured flowers (Figs. 2.6 and 2.7). It flowers from February to April, at which time usually its upper part becomes leafless. Old leaves and flowers are rarely borne together and, therefore, at the time of its flowering, the areas in the forest impart the look of fire-flames (hence the name 'Flame of the Forest'). The tree becomes stunted when its leaves are periodically chopped off for fodder. The colour made from the *kiṁśuka* flowers

Fig. 2.2 Close-up of *Aśvattha* leaves.

Fig. 2.3 Bichrome painting on pottery from Kunal, depicting the peacock and *Pīpala* leaf, ca. 3000-2800 BC.

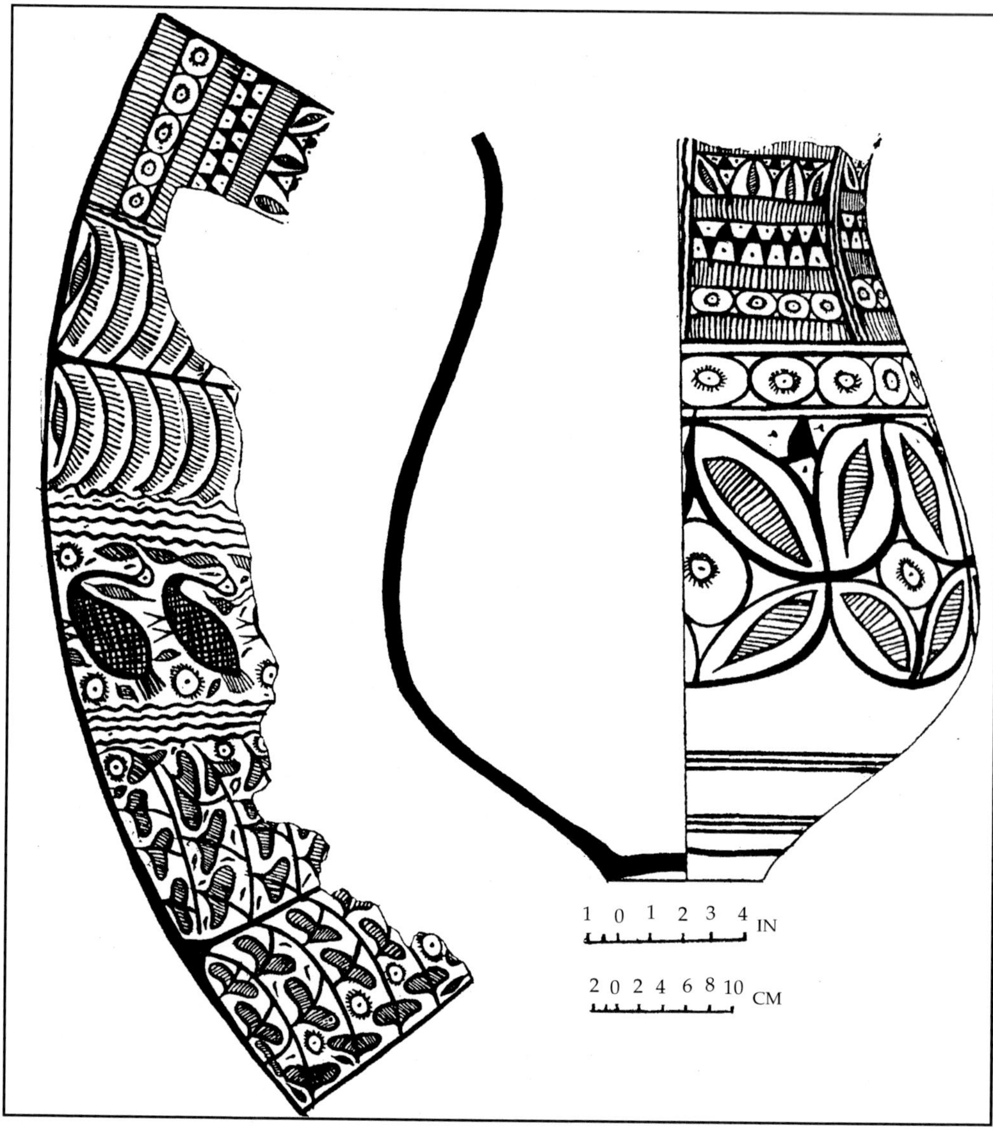

Fig. 2.4 Lothal: Painting on a vase showing *Pīpala* leaf and peacock. Mature Harappan.

Fig. 2.5 Mundigak (Afghanistan): *Pīpala*-leaf motif, painted on pottery, ca. 2700-2500 BC.

is used for dyeing clothes. Also, it is mixed with water and the same is used by Hindus for sprinkling on one another when they celebrate the Holi festival.

There is another species of *Butea*, namely *Butea superba* Roxb., which is of common occurrence in the forests of Avadh (Uttar Pradesh), central India and southern regions (Gamble 1902). It is somewhat smaller than *Butea monosperma* and is horizontally spread out, sometimes becoming even a climber. Flowers and leaves of this species are scarcely distinguishable from those of the other. This is called *palāśa/dhāka* in Hindi, and its broad leaves are preferentially used for making disposable plates and bowls (called respectively *pattal*s and *daunās*), generally used on the occasion of community feasts.

Fig. 2.6 *Kiṁśuka* (*Butea monosperma* Lamk.)

Fig. 2.7 Close-up of *Kiṁśuka* flowers.

In an Indian *materia medica*, namely *Bhāvaprakāśa Nighaṇṭu* (ca. AD 1600), amongst the nine Sanskrit names cited for this tree, *kiṁśuka* and *parṇa* appear as synonyms, and may, therefore, be ascribed to either or both the species of *Butea*. The relevant quote from the *Bhāvaprakāśa Nighaṇṭu* (edited by G.S. Pandey, 1995: 535) is as follows:

Palāśaḥ Kiṁśukaḥ parṇo yajñiyo raktapuṣpakaḥ /

Kṣāraśreṣṭho vātapotho brahmavṛikṣaḥ samidvaraḥ //

As would be seen from hymn ***RV* 10.85.20**, the wood of *kiṁśuka*, along with that of *śalmali* (*Salmalia malabarica*) was used in the making of a chariot.

Sukiṁśukam śalmalim *viśvarūpam hiraṇyavarṇam suvṛitam suchakram /*

Āroha Sūrye amṛitasya lokam syonam patye vahatum kṛiṇuṣva //

Mount this, all-shaped, gold-hued, with strong wheels fashioned of **Kiṁśuka** and **Śalmali**, light-rolling,

Bound for the world of life immortal, Sūryā: make for thy lord a happy bridal journey.

Regarding its geographical distribution, Saraswat states as follows:

It occurs throughout the plains in India, Bangladesh, Myanmar and Sri Lanka, usually in open country in grasslands and is also scattered in mixed forests such as those of Sāla. Usually thrives on black cotton soil, also on salt lands and water-logged places. (Hooker 1874; Gamble 1902; Brandis 1906; Haines 1922; Parker 1918; Stewart 1972). In Pakistan, *Butea* trees run wild in the foothill zone from Rawalpindi District eastwards, up to a height of 1200 metres, and also cultivated in plains (Stewart 1972).

(iii) *Khadira* (*Acacia catechu* Willd.)

Called *khaira* in Hindi and known for the hardness of its wood, this tree (Fig. 2.8) was also recognized by the Ṛigvedic people as being fairly strong and was therefore used for making chariots, as may be gleaned from the following hymn.

***RV* 3.53.19:**

Abhi vyayasva ***khadirasya*** *sāramojo dhehi spandane* ***śiṁśapāyām*** /
Akṣa vīḷo vīḷita vīḷayasva mā yāmādasmādava jīhipo naḥ //

Enclose thee in the heart of **Khayar** timber, in the car wrought of **Śiṁśapā** put firmness.

Show thyself strong, O Axle, fixed and strengthened; throw us not from the car whereon we travel.

In the context of this tree, Saraswat states:

It is found only in India, Pakistan and Myanmar. It occurs throughout the sub-Himalayan tracts from the Indus eastwards, ascending to 900 metres; also in Madhya Pradesh, Bihar and Orissa. The *khaira* tree found in Assam and Bengal is regarded to be a distinct variety (*Acacia catechu* var. *catechuoides*), which occurs throughout western and southern India (Gamble 1902; Brandis 1906; Ramesh Rao and Purkayastha 1972). In Pakistan, it is found scattered on the foothills in western Himalayas, ascending up to 1200 metres in Peshawar, Rawalpindi and Swat Divisions (Stewart 1972).

To the foregoing, Saraswat further adds:

Belonging to the same genus, viz. *Acacia*, is the *babūla* (*Acacia nilotica*). It is a tree of arid and dry regions. Charcoal pieces of *Acacia* wood have been found in the Early Harappan Period at Kunal and in the Mature Harappan Period at Banawali, Rohira and Sanghol. Since the anatomical features of both these species are identical, the problem of specific identification cannot be easily resolved on

Fig. 2.8 *Khadira* (*Acacia catechu* Willd.)

botanical grounds. At present, the *babūla* tree is predominantly found in the arid regions of Rajasthan, Panjab and Haryana. In the past when the mighty Sarasvatī was the lifeline of the Harappans in this region, circumstantial evidence indicates that the *khadira* (*Acacia catechu*) was of common occurrence in the aforesaid regions.

(iv) *Nyagrodha (Ficus benghalensis* L.)

Also known as *banyan* in English and *baragada* in Hindi, it is a very large tree, with spread-out stout branches from which descend many aerial roots down to the ground level (Fig. 2.9). These roots, in turn, grow into trunks. In the Botanical Gardens at Kolkata there is a famous banyan tree which is stated to be very old. While its main trunk has a circumference of over 12 m, the subsidiary trunks, developed from the descending roots and numbering about 230, have circumferences varying from 1.8 m to 3 m. The banyan tree also has a religious association amongst the Hindus. Womenfolk perform special worship

Fig. 2.9 *Nyagrodha* (*Ficus benghalensis L.*)

at this tree on the occasion of what is known as *Baragadī Amāvasyā.*

As in the case of the *pīpala,* birds are fond of eating the fruits of the *baragada* as well. After eating the fruits, when the birds sit on other trees and pass excreta, the *baragada*-seeds germinate on those trees, and the same are damaged in the course of time. In a like manner, *baragada* and *pīpala* trees also grow on buildings and affect them badly. Seeds once rooted in the crevices are almost difficult to extricate.

Although not mentioned by name, the *nyagrodha* seems to be implied in Hymn **1.24.7** of the *Ṛigveda,* since its characteristic features are mentioned therein. The hymn reads as follows:

Abudhne rājā Varuṇo vanasyordhvam stūpam dadate pūtadakṣaḥ /
nīchīnāḥ sthurupari budhna eṣāmasme antarnihitāḥ ketavaḥ syuḥ /

Varuṇa, King, of hallowed might, sustaineth erect the Tree's stem[2] in the baseless region.

Its rays, whose root is high above, stream downward. Deep may they sink within us, and be hidden.

The Ṛigvedic description being what it is, no indication is given therein of the use to which the wood of this tree was put. However, later texts, for example the *Taittirīya Saṁhitā* 7.4.12.1, show that its wood was used for making sacrificial bowls (*chamasa*).

The following is the distribution-pattern of this tree, as communicated by Saraswat:

> The tree runs wild in the sub-Himalayan forests from Peshawar to Assam; deciduous forests of Bihar, Chhota Nagpur, Orissa, Madhya Pradesh and South India; also, but less common, in evergreen forests; low country of Sri Lanka (Gamble 1902; Brandis 1906; Hooker 1878; Talbot 1911). The banyan tree is planted in the plains of Sind, Punjab and Baluchistan and is reported to run wild in the foothill zone from the Indus eastward, often on cliffs up to 900 m (Stewart 1972). Cultivated in the Jalalabad region of Afghanistan (Kitamura 1960).

---

2. In the Sanskrit text, the word used is *stūpa* which does not mean 'stem', as translated by Griffith. It suggests that the upper profile of the tree, which looks like a semi-circular canopy, resembles a *stūpa.*

Fig. 2.10 *Vibhīdaka* (*Terminalia bellerica* Roxb.)

(v) *Vibhīdaka* / *Vibhītaka* (*Terminalia bellerica* Roxb.)

Of the two Sanskrit spellings of the word given above, the former is the older one, occurring as it does in the *Ṛigveda*. The other is more common in the later texts. In Hindi it is called *baherā* (Fig. 2.10).

As indicated by the Ṛigvedic references, the nuts of this tree were used for dicing or maybe the dice were made out of its wood. Later Vedic texts show that its wood was used also in sacrificial fire. The following are the Ṛigvedic reference:

***RV* 7.86.6:**

Na sa svo dakṣo varuṇa dhrutiḥ sā surā manyur ***vibhīdako*** *achittiḥ* /

Asti jyāyān kanīyasa upāre svapnaśchanedanṛitasya prayotā //

Not our own will betrayed us, but seduction, thoughtlessness, Varuṇa!, wine, **dice**, or anger.

The old is near to lead astray the younger: even sleep removeth not all evil-doing.

***RV* 10.34.1:**

Prāvepā mā bṛihato mādayanti pravātejā iriṇe varvṛitānāḥ /

Somasyeva Maujavatasya bhakṣo ***vibhīdako*** *jāgṛivirmahyamachchhān* //

Sprung from tall trees on windy heights, these rollers transport me as they turn upon the table.

Dearer to me the **die** [made of *Vibhīdaka* tree] that never slumbers than the deep drought of Mūjavān's own Soma.

Says Saraswat:

This tree occurs throughout the forests of India, Myanmar and Sri Lanka, below elevations of 900 m, but not in the dry and arid country of Sindh and Rajputana (Gamble 1902; Hooker 1874). Also found in Malaya (Hooker 1874). In Pakistan, found from Panjar, near Kahuta, down to the Jhelum river and diffusing towards Udhampur and Jammu (Stewart 1972).

(vi) *Śalmali* (*Bombax ceiba* L., Syn. *Salmalia malabarica* [DC.] Schott.)

Known in Hindi as *senwala* or *senbhala*, this large tree bears flowers which, on opening, cast away very fine, silk-soft white floss (Figs. 2.11 and 2.12). The latter is often used in stuffing quilts and pillows.

In the *Ṛigveda*, this tree occurs in two different contexts. In one, it is stated that some kind of poison is formed on it. In the other, it is indicated that its wood, along with that of the *kiṁśuka*, was used for manufacturing chariots. The respective hymns are as follows:

***RV* 7.50.3:**

Yachchhalmalau *bhavati yannadīṣu yadoṣadhībhyaḥ pari jāyate viṣam /*

Viśve devā niritastat suvantu mā mām padyena rapasā vidat tsaruḥ //

The poison that is formed upon the **Śalmali,** that which is found in streams, that which the plants produce,

All this may all the Gods banish and drive away: let not the winding worm touch me and wound my foot.

Hymn ***RV* 10.85.20**, which refers to the construction of the chariot from the wood of the *śalmali* tree, has already been quoted in the context of the *kiṁśuka* tree (see p. 11) and is, therefore, not repeated here.

Fig. 2.11 *Śalmali* (*Bombax ceiba* L.)

As regards the spatial distribution of the *śalmali* tree, Saraswat observes:

> It occurs throughout India, Myanmar and Sri Lanka; also in Java and Sumatra (Gamble 1902; Brandis 1906; Hooker 1874). Also in the sub-Himalayan tract from Hazara (Pakistan) eastward, ascending up to 900 m (Stewart 1972).

It is interesting to note that charcoal pieces of the *śalmali* wood have been found in the Harappan levels at Rohira.

(vii) *Śiṁśapā* (*Dalbergia sissoo* Roxb.)

This tree is known as *śīśama* in Hindi (Fig. 2.13). Because of its intrinsic strength, its wood was used by the Ṛigvedic people for making chariots.

At present too, the wood of *Dalbergia sissoo* is regarded as the best for all such purposes where strength and elasticity are required. For felloes and naves of wheels and for frames of carriages, it is unsurpassed by any other wood, owing to its fine seasoning and durability. In this context, it may not be out of place to quote from Gamble (1902) wherein he refers to a report by a senior army officer, as follows:

> The wheels of our ordnance carriages have never failed, however arduous or lengthened the service has been on which they have been employed, of which no more striking example can be furnished than the campaign in Afghanistan, about the most trying country in the world for wheels. Some of our batteries served

Fig. 2.12 Close-up of the flowery part of the *Śalmali.*

> throughout the campaign, went to Bameen (sic) and even to Hindoo Kush (sic), and came back again to India without a breakdown, while Royal Artillery wheels, built out of the very best materials Woolwich could produce, specially for Indian service, almost fell to pieces after a few months' exposure and service on the plains of India.

The hymn, viz. ***RV* 3.53.19**, in which the *śiṁśapā* tree has been referred to, has already been quoted (p. 12) in the context of the *khadira* tree, together with which it was used, as just stated, for the construction of chariots. There is little point in repeating the hymn here.

In the context of the distribution-pattern of *Dalbergia sissoo*, Saraswat states:

> In India, this tree occurs throughout the sub-Himalayan tracts from the Indus to Assam, ascending usually up to 900-m elevation, but sometimes even up to 1,500 m in central Himalayas; often extending far into the plains (Gamble 1902; Brandis 1906; Ramesh Rao and Purkayastha 1972). It also occurs in Baluchistan (Pakistan)

Fig. 2.13 *Śiṁśapā* (*Dalbergia sissoo* Roxb.) in the background. *Muñja* grass (*Saccharum bengalense* Retz.) in the foreground.

and parts of Afghanistan (Hooker 1874; Stewart 1972). It is also reported from parts of south-eastern Iran, contiguous to southern Afghanistan. However, no occurrence of *Dalbergia sissoo* is recorded in the region of former USSR in Central Asia. There are two different species of *Dalbergia,* namely *Dalbergia simplex* and *Dalbergia sewardina*, of sporadic distribution in this region (Komarov 1948).

Charcoal pieces of the wood of *Dalbergia sissoo* have been discovered in the Early Harappan levels at Kunal and in the Mature Harappan levels at Banawali and Rohira. Further, it is interesting to note that in the Mesopotamian records of the third millennium BC, the *sissoo* wood is often referred to as an item of import, the sources being Maggan and Meluḥḥa. These regions are generally equated respectively with the the coastal Makran parts of Iran and Pakistan and with the area covered by the Harappan Civilization in India and Pakistan. (Cf. Lal 1997: 194-97).

B. PLANTS AND GRASSES

(i) *Urvāruka* (*Cucumis melo* L. var. *utilissimus* Roxb. Duthie et Fuller, Syn. *Cucumis utilissimus* Roxb.)

Cucumis melo covers a large number of varieties. These freely intercross and produce the variable form in wild and cultivated state. The species *Cucumis utilissimus* has currently been given the status of a variety of cultivated *Cucumis melo*. There are three varieties, *viz.*: *Cucumis melo* L. var. *melo* (*kharbūjā* in Hindī), *C. melo* L. var. *momordica* (*phūta* in Hindī) and *C. melo* L. var. *utilissimus* (*kakaṛī* in Hindī). These are grown mainly in summer and end up by the beginning of the rainy season.

The Ṛigvedic hymn (***RV* 7.59.12**) is philosophically very significant. The devotee prays that he/she may be released from the bondage of death, just as the *urvāruka* fruit is released from the body of the plant (as soon as it is ripe) and, the prayer goes on to add that he/she may, thereafter, be joined to (i.e. granted) immortality. The hymn runs as follows:

tryambakam yajāmahe sugandhim puṣṭivardhanam /
urvārukamiva *bandhanānmṛityormukṣīya māmṛitāt //*

Tryambaka we worship, sweet augmenter of prosperity.
As from its stem the **cucumber**, so may I be released from death, not reft of immortality.

As regards the distribution-pattern of this plant, Saraswat states, as follows:

> *Cucumis melo* has been reported to occur wild in India, Baluchistan and tropical Africa (Hooker, 1874; De Candolle, 1886; Duthie, 1903-29; Haines, 1922; Stewart, 1972). *Cucumis melo* from which the groups of cultivated forms of *kharbūjā*, *phūṭa* and *kakaṛī* have developed, is regarded to have run in truly wild forms in eastern tropical Africa, south of the Sahara. Once domesticated, however, it exploded into a number of cultivars, particularly in India, which may thus be regarded as a secondary centre (Whitaker and Bemis, 1976). Wild forms from India are more likely to be feral escapes from local cultivars.
>
> *Cucumis melo* var. *utilissimus* (*kakaṛī*) implied for *urvāruka*, along with *c. melo* var. *momordica* (*phūṭa*), is cultivated rather extensively in the areas of upper and middle Gangetic plain, north-western India and Pakistan. Muskmelon (*C. melo* var. *melo*), *kharbūjā* in Hindi, is more esteemed and cultivated throughout India, chiefly on sandy beds and margins of rivers.

(ii) *Darbha* (*Imperata cylindrica* [L.] P. Beauv. -Syn. *Saccharum cylindricum* [L.] Lamk.)

The name occurs along with five others which, judging from the context, also seem to be grasses of one kind or another. The relevant Ṛigvedic verse (**1.191.3**) is as follows:

Śarāsaḥ Kuśarāso ***Darbhāsaḥ*** *Sairyā uta /*

Mauñjā adṛiṣṭā Vairiṇāḥ sarve sākam nyalipsata //

Śara grass, **Darbha**, Kuśara, and Sairya, Muñja, Viraṇa,

Where all these creatures dwell unseen, with poison have infected me.

From the context it is clear that these grasses are infested with poisonous insects like snakes, etc. Such luxuriant growth of grasses sheltering poisonous reptiles indicates a hot and humid climate. About the habitat of the *darbha* grass Saraswat states as follows:

> An aggressively rhizomatous perennial grass of common occurrence in almost all warm countries. It has three varieties. *I. cylindrica* var. *cylindrica* is found in the Mediterranean region, extending eastward to Iraq, Iran, Afghanistan and parts of Pakistan and also arid stretches of the former USSR. *I. cylindrica* var. *latifolia* is of rather limited distribution in swampy places of Garhwal, parts of Panjab and Jammu & Kashmir. Variety *major* of *I. cylindrica* covers a very large area throughout the hotter parts of India and Pakistan, and extends up to Australia on the east and South Africa on the west (Hooker 1880; Bor 1960; Cope 1982). In eastern parts of India and Bangladesh, any bare areas, such as slips in the hills and abandoned fields, are rapidly taken possession of. The succulent rhizomes are very tenacious and a complete plant will grow from even a small piece, so that this grass is extremely difficult to eradicate.

(iii) *Pākadūrvā (Cynodon dactylon* [L.] Pers., Syn. *Panicum dactylon* L.)

A grass, widely creeping and forming matted tuft with short ascending branches (Fig. 2.14). In the funeral hymn of the *Ṛigveda* (**10.16.13**) it is mentioned along with *kiyāmbu* and *vyalkaśā* – as plants/grasses that grow/are grown on the place where the dead body is cremated.

Yam tvamagne samadahastamu nirvāpayā punaḥ /

kiyāmbvatra rohatu ***pākadūrvā*** *vyalkaśā //*

Cool, Agni, and again refresh the spot which thou hast scorched and burnt.

Here let the water-lily grow, and tender grass and leafy herbs.[3]

---

3. Griffith's translation of the second line of the foregoing Sanskrit verse is a very generalized one, ignoring the names of the individual plants/grasses.

Fig. 2.14 *Pākadūrvā* (*Cynodon dactylon* [L.] Pers.)

It seems likely that the *pākadūrvā* grass was not different from the *dūrvā* grass mentioned in certain verses. For example, ***RV* 10.142.8:**

Āyane te parāyaṇe ***dūrvā*** *rohantu puṣpiṇīḥ /*
hradāścha puṇḍarīkāṇi samudrasya gṛihā ime //

On thy way hitherward and hence (O Agni) let flowery **Dūrvā grass** spring up.

Let there be lakes with lotus blooms. These are the mansions of the flood.

In ***RV* 10.134.5,** there occurs a simile which seems to suggest that the ears of this particulars kind of grass lay horizontal with the stem:

ava svedā ivābhito viṣvak patantu didyavaḥ /
dūrvāyā iva tantavo *vyasmadetu durmatirdevī janitryajījanad bhadrā janitryajījanat //*

Around, on every side like drops of sweat let lightning-flashes fall.

Let all malevolence pass away from us like threads of **Dūrvā grass.**

As regards the distribution-pattern of ***pākadūrvā/dūrvā***, Sarswat says: 'It is commonly found in the plains of India, Pakistan, Myanmar and Sri Lanka (Hooker 1880; Bor 1960; Cope 1982). Stewart (1972) reports that in Pakistan it is even cultivated, for fodder.'

(iv) *Muñja (Saccharum bengalense* Retz. Syn. *S. munja* Roxb.)

It is a very large erect grass, sometimes rising to a height of about 6 m (Fig. 2.15). It grows on alluvial sandy soil, on banks of streams not subject to water-logging. Genus *Saccharum* has about 35 species of clump-forming rhizomatous grasses with cane-like stems. All these are widely distributed in tropical and warm temperate regions of the Old World (Hyam and Pankhurst 1995).

The species are valuable for fibers (*mūñja*), extracted from the upper leaf-sheaths of the culms, which are employed in making ropes and mats, baskets, etc. Fibres are resistant to attack by ants. Leaves are used for thatching, thicker stems for screens, tools, baskets, etc. and upper portions of flowering culms for making winnowing trays and even bullock-cart covers.

In the *Ṛigveda* (**1.191.3**), it has been mentioned along with other kinds of grasses as a lurking place for venomous creatures:

Śarāsaḥ kuśarāso darbhāsaḥ sairyā uta /

***Mauñjā** adṛiṣṭā vairiṇāḥ sarve sākam nyalipsata //*

Śara grass, Darbha, Kuśara and Sairya, **Muñja**, Vīraṇa,

Where all these creatures dwell unseen, with poison have infected me.

In ***RV* 1.161.8**, however, it is mentioned in a different context, most likely as a material used for filtering the Soma:

*Idamudakam pibatetyabravītanedam vā ghā pibatā **muñjanejanam** /*

saudhanvanā yadi tanneva haryatha tṛitīye ghā savane mādayādhvai //

Drink ye this water, were the words ye spake to them; or drink ye this, the **rinsing of the Muñja-grass**.

Fig. 2.15 *Muñja* (*Saccharum bengalense* Retz. Syn. *S. munja* Roxb.)

If ye approve not even this, Sudhanvan's sons, then at the third libation gladden ye yourselves.

On a small piece of mud-plaster from the Early Harappan levels at Kunal (ca. 3000-2800 BC) have been found minute remains of leaf-epidermis of the *muñja* grass, along with a few charcoal pieces of bamboo (*Bambusa* Sp.). These are regarded by Saraswat and Pokharia (2003) to have been used for making some wattle-and-daub structure at the site.

About the distribution of the *muñja* grass, Saraswat states: '*Saccharum bengalense (Muñja)* is of common occurrence throughout the plains and low hills of India, Pakistan, Bangladesh, Afghanistan, Sri Lanka and parts of China (Hooker 1880; Bor 1960; Kitamura 1960).'

(v) *Śīpāla* (*Blyxa octandra* Planch. Syn. *B. roxburghii* Rich.)

This plant grows in the still waters of tanks and small lakes and is often tossed about by winds. It has been referred to in the *Ṛigveda* **(10.68.5)** in the following context:

> *apa jyotiṣā tamo antarikṣādudnaḥ* ***śīpālamiva vāta ājat*** /
> *bṛihaspatiranumṛiśyā valasyābhramiva vāta ā chakra ā gāḥ* //
>
> Forth from mid air with light he drave the darkness, **as the gale blows a lily[4] from the river.**
>
> Like the wind grasping at the cloud of Vala, Bṛihaspati gathered to himself the cattle.

About this plant Saraswat writes as follows:

> Submerged, tufted and stemless plant, common in still water. Distributed widely in **tropical Asia and Australia.** In India, it is of common occurrence in Uttar Pradesh, Bihar, Bengal, Khasia Hills (Meghalaya), Orissa and Tamil Nadu (Hooker 1878; Duthie 1903-29; Haines 1924). The plant in north Indian regions like Panjab and Kashmir and in Sindh (Pakistan) is represented by another species – *Blyxa oryzetorum* (Dcne.) Hook. f. (Stewart 1972).

(vi) *Soma* (Species of *Sarcostemma* ?/ *Ephedra* L. ?)

No discussion on the Ṛigvedic flora will be complete without a reference to *Soma*, although, unfortunately, it has not been possible to identify this plant with full certainty. It is for this reason that a question-mark has been put in the heading above.

The plant played such a vital role in the life of the people that in the *Ṛigveda* a whole *maṇḍala* (viz. 9), comprising as many as 114 hymns, was devoted to it. Besides, there are many hymns in other *maṇḍala*s as well. Its juice was used in Vedic sacrifices; hence its great importance. To extract the juice from it, the plant was pounded with stones or in a mortar (*ulūkhala*) with a pestle (*manthā*), as may be noted from the following verses:

---

4. This is a generalized rendering of *śīpāla*.

***RV* 1.28.1:**

yatra ***grāvā pṛithubudhna ūrdhvo bhavati sotave*** */*

ulūkhalasutānām*avedvindra jalgulaḥ //*

There where the **broad-based stone is raised on high to press the juices out,**

O Indra, drink with eager thirst the **droppings which the mortar sheds.**

***RV* 1.28.4:**

yatra ***manthām vibadhnate*** *raśmīn yamitavā iva /*

ulūkhalasutānām*avedvindra jalgulaḥ //*

Where, as with reigns to guide a horse, they **bind the churning-staff with cords,**

O Indra, drink with eager thirst the **droppings which the mortar sheds.**

From verses like the following, it is learnt that this plant grew on mountains, amongst which the is specially mentioned.

***RV* 10.34.1:**

prāvepā mā bṛihato mādayanti pravātejā iriṇe varvṛitānāḥ /

Somasyeva Maujavatasya *bhakṣo vibhīdako jāgṛivirmahyamachchhān //*

Sprung from tall trees on windy heights, these rollers transport me as they turn upon the table.

Dearer to me the die that never slumbers than the deep draught of **Mūjavān's own Soma.**

Although there is a debate about the identification of the Mūjavant, a reference (though later) in the *Mahābhārata* (10.17.26; 14.8.1) would seem to suggest that it may have been a range in the Himalayas. (The actual word occurring in the *Mahābhārata* is Muñjavant which Yaska takes to denote the Vedic Mūjavant.) There could be yet another way of looking at the identification of the Mūjavant mountain. In the *Atharvaveda* (5.22.5, 7, etc.) Mūjavant is mentioned as the name of a people, along with Gandhāris and Balhikas, from which it would appear that the Mūjavant people derived their identity from the mountain of that name. This would place the Mūjavant somewhere in eastern Afghanistan. Be that as it may, the important point to note is that the source of the *Soma* plant was somewhere in the north-western part of pre-Partition India and the adjacent region of Afghanistan.

It is interesting to note that the *Soma,* in its modified form *Haoma* (*sa* changing to *ha*) is referred to in the *Avesta,* the sacred book of the Parsis. Whatever be the dates of the *Avesta* and the *Ṛigveda* in terms of absolute BCs, it has been acknowledged on all hands that the former is later than the latter. It is, thus, evident that when the ancient Parsis parted their way from the Vedic Aryans they took the *Soma* tradition with them.

Before closing this discussion on the *Soma* plant, it may not be out of place to put on record some information which I gathered about thirty years ago during a visit to Afghanistan to participate in a conference. Some local scholars informed me that in the north-eastern part of that country there lives a tribe called Kafir. (The term *Kafir* is used by Muslims for unbelievers, which shows that this tribe does not subscribe to Islam but follows some other religion.) The people of this tribe, the scholars told me, extract the juice of a plant which is even now called the *haoma.* They do so on ritualistic occasions and find the drink exhilarating. I very much wanted to visit this area to check things up for myself but for various reasons this could not be done. Seeing that the relations between Afghanistan and India are now amicable, recently at a meeting of the Central Advisory Board of Archeology, held in New Delhi, I put forward a proposal for such an investigation by a joint Indo-Afghan team and am anxiously awaiting the outcome.

CHAPTER III

THE FAUNA

As in the case of the flora, in the fauna also two sub-divisions are being made, dealing respectively with: (A) Animals and (B) Birds.

A. ANIMALS

(i) *Uṣṭra* (*Camelus dromedarius* and *Camelus bactrianus*)

The camel (*uṣṭra*) has been referred to in about half-a-dozen verses in the *Ṛigveda*. In ***RV* 10.106.2** there occurs a word, *uṣṭi*, which in all likelihood also denotes a camel. Some of these verses are reproduced below, along with Griffith's translation thereof:

***RV* 1.138.2:**

Pra hi tvā Puṣannajiram na yāmani stomebhiḥ kṛiṇva ṛiṇavo yathā mṛidha ***uṣṭro*** *na pīparo mṛidhaḥ /*

huve yat tvā mayobhuvam devam sakhyāya martyaḥ /

asmākamāṅgūṣān dyumninaskṛidhi vājeṣu dyumninaskṛidhi //

Thee, then, O Pūṣan, like a swift one on his way, I urge with lauds that thou mayst make the foemen flee, drive, **camel-like,** our foes afar.

As I, a man, call thee, a God, giver of bliss, to be my Friend,

So make our loudly-chanted praises glorious, in battles make them glorious.

RV 8.5.37:

tā me aśvinā sanīnām vidyātam navānām /

*yathā chichchaidyaḥ kaśuḥ śatam**uṣṭrānām** dadat sahasrā daśa gonām //*

As such, O Aśvins, find for me my share of new-presented gifts,

As Kaśu, Chedi's son, gave me a hundred head of buffaloes,[1] and ten thousand kine.

RV 8.6.48:

*udānaṭ kakuho divam**uṣṭrāñ** chaturyujo dadat /*

śravasā yādvam janam //

Kakuha hath reached up to heaven, bestowing buffaloes[2] yoked in fours, and matched in fame the Yādavas.

RV 8.46.22:

*ṣaṣṭim sahasrāśvyasyāyutāsanam**uṣṭrānām** viṁśatim śatā /*

daśa śyāvīnām śatā daśa tryaruṣīṇām daśa gavām sahasrā //

Steeds sixty thousand and ten thousand kine, and twenty hundred **camels** I obtained;

Ten hundred brown in hue, and other ten red in three spots: in all, ten thousand kine.

RV 8.46.31:

*adha yachchārathe gaṇe śatam**uṣṭrām** achikradat /*

adha śvitneṣu viṁśatim śatā //

And in the grazing herd he made a hundred **camels** bleat for me,

And twenty hundred mid the white.

1. Griffith here wrongly translates the word *'uṣṭra'* as 'buffalo'. It should have been 'camel', as in his own translation of the previous verse.
2. Here again Griffith has mis-translated the word *'uṣṭra'* as 'buffalo'.

Fig. 3.1 One-humped camel (*Camelus dromedarius*).

There are two main species of the camel. Of these, the more widely spread is *Camelus dromedarius*, which is distinguished by its single hump (Fig. 3.1). It is found mainly from Arabia on the west to India on the east. Further west, camels are also met with in northern Africa; and in the east one may see them in Australia as well where these were introduced in modern times to help in the exploration of the vast extent of that country. The other species, known as *Camelus bactrianus* and characterized by its two humps (Fig. 3.2), is native to Afghanistan, Mongolia and extends to even China. Whereas, the latter species, besides being domesticated, may still be found in the wild in the Gobi desert, of the former no wild habitats appear to have been identified. Besides its two humps, another feature that distinguishes the *Camelus bactrianus* from its counterpart, is the presence of thick long hair over its neck, humps and even on some other parts of the body, which keep it warm and protect it from the cold climate of the terrain. Both the species are herbivores, i.e. feed on grass,

Fig. 3.2 Two-humped camel (*Camelus bactrianus*).

foliage of trees and bushes, etc.

The camel is used for transporting goods as well as for riding. It can run as fast as 15 km per hour and can score 150 km in a day. Because of the special kind of make-up of its paws, the camel can easily walk on soft desert sand which is very difficult to be negotiated even by wheeled vehicles. It is for this reason that the camel is nicknamed as 'the ship of the desert'. Certain other physical characteristic features make the camel perfectly fit to cope with the hard desert conditions, in particular its heat, dryness and inadequate water-supply. It can store plenty of water in the lining of its stomach. Its hump not only provides protection from the scorching summer heat but also stores fat which is metabolized to provide energy.

Besides being used for transportation and ferrying across vast deserts, the camel has many other economic uses. For example, its wool is used for making clothes and blankets, while its skin is used for preparing shoes, water-bags, etc.

Though not portrayed on their seals, the Harappan people did domesticate this animal, as evidenced by the discovery of its skeletal remains from many Harappan sites, such as Mohenjo-daro, Harappa, Kalibangan, Surkotada, etc. This takes the antiquity of the use of this animal at least to the third millennium BC.

(ii) *Gaura* (*Bos gaurus* H. Smith)

The *gaura*, sometimes with *gavaya*, has been mentioned in about half-a-dozen verses of the *Ṛigveda*. The female, *gaurī*, too has been referred to in at least three verses. In all such cases it is evident that the domesticated form of the animal was intended, since, though not in the *Ṛigveda* itself, elsewhere such terms as *āraṇya gaura* and *gaura-mṛiga* (e.g. *Vājasaneyī Saṁhitā*, 13.48 and 24.32) occur referring to the animal in its wild state. Some of the verses of the *Ṛigveda* are quoted below:

***RV* 1.16.5:**

semam naḥ stomamā gahyupedam savanam sutam /
***gauro** na tṛiṣitaḥ piba //*

Come thou to this our song of praise, to the libation poured for thee:
Drink of it like a stag[3] athirst.

***RV* 4.21.8:**

vi yad varānsi parvatasya vṛiṇve payobhirjinve apām javānsi /
*vidad **gaura**sya gavayasya gohe yadī vājāya suddhyo vahanti //*

When he unbars the spaces of the mountains, and quickens with his floods the water-torrents,

---

3. In the translation of the verse that follows, Griffith uses the words 'buffalo' and 'wild ox' for the Sanskrit words *'gaura'* and *'gavaya'*. It is thus clear that Griffith was never very clear in his mind about the identification of the animal called *'gaura'*. This word refers to a species of ox, called *Bos gaurus*. For the details relating to this animal please see pp. 35-36 and Fig. 3.3.

He finds in lair the buffalo and wild ox when the wise lead him on to vigorous exploit.

***RV* 4.58.2:**

vayam nāma pra bravāmā ghṛitasyāsmin yajñe dhārayāmā namobhiḥ /
upa brahmā śṛiṇavachchhasyamānam chatuḥ śṛiṅgovamīd ***gaura*** *etat //*

Let us declare aloud the name of Ghṛita, and at this sacrifice hold it up with homage.

So let the Brahman hear the praise we utter. This hath the four-horned buffalo[4] emitted.

***RV* 5.78.2:**

aśvinā hariṇāviva ***gaurā****vivānu yavasam /*
haṁsāviva patata mā sutāṁ upa //

O Aśvins, like a pair of deer, like two **wild cattle** to the mead:
Fly hither like two swans unto the juice we shed.

Here, and in the verse that follows, Griffith translates the word *'gaura'* as 'wild cattle'.

***RV* 7.69.6:**

narā ***gaureva*** *vidyutam tṛiṣāṇāsmākamadya savanopa yātam /*
purutrā hi vām matibhirhavante mā vāmanye ni yaman devayantaḥ //

Like the **wild cattle** thirsty for the lightening, Heroes, come nigh this day to our libations.

Men call on you with hymns in many places, but not let other worshippers detain you.

---

4. Please see the foot-note on p. 33. It has to be further stated that there already exists a separate word, viz. *mahiṣa*, for buffalo (see below, p. 36 ff.). So, how can both *gaura* and *mahiṣa* be identical?

There are many more verses referring to the *gaura*, but not much useful purpose would be served by quoting them. However, we do give below a couple of verses wherein the word *gaurī* occurs signifying the female member of the *gaura* family.

***RV* 1.84.10:**

svādoritthā viṣūvato madhvaḥ pibanti ***gauryaḥ*** */*

yā indreṇa sayāvarīrvṛiṣṇā madanti śobhase vasvīranu svarājyam //

The juice of Soma thus diffused, sweet to the taste, the bright **cows** drink,

Who for the sake of splendour close to mighty Indra's side rejoice, good in their own supremacy.

***RV* 9.12.3:**

madachyut kṣeti sādane sindhorūrmā vipaśchit /

somo ***gaurī*** *adhi śritaḥ //*

In the stream's wave wise Soma dwells, distilling rapture, in his seat,

Resting upon a **wild cow's** hide.

We may now turn our attention to the physical and other aspects of the ***gaura*** **(*Bos gaurus*)**. It is a strong and heavily built animal, characterized by its prominent muscular ridge above its shoulders, which slopes down as it reaches the middle of the back. Yet another feature is the massiveness of the head, which has a bulging ridge between its thick and incurved horns (Fig. 3.3). Its colour varies from reddish brown to almost shiny black. However, the lower parts of the legs are usually cream-coloured with rump patches. Its height, at the shoulder-ridge, can sometimes reach up to 2 m, while the weight in the case of males can go even up to 1000 kg.

According to the 1995 Asian Wild Cattle Conservation Assessment and Management Plan, there are three subspecies of the wild gaur: (i) *Bos gaurus gaurus* of India and Nepal: (ii) *Bos gaurus hubbacki* of Thailand and Malaysia; and (iii) *Bos gaurus laosiensis* inhabiting the stretch from Myanmar to China. The natural habitat of the gaur is primarily tropical monsoon and dry forests, usually below the altitude of 1500 and 1800 m. Its feeds mainly on dry grasses,

Fig. 3.3 *Gaura* (*Bos gaurus*).

bushes and trees. It is important to note that the **majority of its present estimated population, which may be anywhere between 15,000 and 30,000, is in India. In sum, it is essentially an animal of South and South-East Asia, having little to do with Europe or western and central Asia.**

(iii) *Mahiṣa* (Water buffalo, *Bubalus bubalis* L./ *Bubalus arnee*)

This animal is mentioned in the *Ṛigveda* at a number of places, some of which are quoted below:

***RV* 5.29.7:**

sakhā sakhye apachat tūyamagnirasya kratvā ***mahiṣā*** *trī śatāni /*
trī sākamindro manuṣaḥ sarānsi sutam pibad vṛitrahatyāya somam//

As friend to aid a friend, Agni dressed quickly three hundred **buffaloes**, even as he willed it.

And Indra, from man's gift, for Vṛitra's slaughter, drank off at once three lakes of pressed-out Soma.

***RV* 8.12.8:**

yadi pravṛiddha satpate sahasram ***mahiṣāṁ*** *aghaḥ /*
ādit ta indriyam mahi pra vāvṛidhe //

When, Mighty Lord of Heroes, thou didst eat a thousand **buffaloes**,
Then grew and waxed exceeding great thine Indra-power.

***RV* 9.87.7:**

eṣa suvānaḥ pari somaḥ pavitre sargo na sṛiṣṭo adadhāvadarvā /
tigme śiśāno ***mahiṣo*** *na śṛiṅge gāvyannabhi śūro na satvā //*

This Soma, pressed into the cleansing filter, hath run as 'twere a host let loose, the Courser;

Like a strong bull[5] who whets his horns keen-pointed, like a brave warrior in the fray for cattle.

While dealing with the word *'mahiṣa'* it needs to be added that sometimes it occurs along with the word *'mṛiga'*. It is necessary to realize the implication of such a combination. And here we quote a couple of such references.

***RV* 9.92.6:**

pari sadmeva paśumānti hotā rājā na satyaḥ samitīriyānaḥ /
somaḥ punānaḥ kalaśāṁ ayāsīt sīdan ***mṛigo na mahiṣo vaneṣu***//

As the priest seeks the station rich in cattle, like a true King who goes to great assemblies,

5. This translation by Griffith, when compared with his own translation of verses ***RV* 5.29.7** and ***RV* 8.12.8** (above), and of verse ***RV* 4.58.2** (p. 34) would show that Griffith keeps on mixing up the bull, buffalo and *Bos gaurus*.

Soma hath sought the beakers while they cleansed him, and like a ***wild*** **bull,**[6] in the wood hath settled.

***RV* 10.123.4:**

jānanto rūpamakṛipanta viprā ***mṛigasya ghoṣam mahiṣasya*** *hi gman /*
ṛitena yanto adhi sindhumasthurvidadgandharvo amṛitāni nāma //

Knowing his form, the sages yearned to meet him: they have come nigh to hear the ***wild*** **bull's**[6] bellow.

Performing sacrifice they reached the river: for the Gandharva found the immortal waters.

The water buffalo (named as *Bubalus arnee* in the wild form and as *Bubalus bubalis* in its domesticated status) is a sturdy animal, measuring 1.5-1.8 m in height at the shoulder-level and stretching up to 3 m in length (Fig. 3.4). Usually black in colour, the body is sparsely overlain with long and coarse hair. It has long narrow face, with a pronounced tuft of hair on the upper part of the forehead. Its horns are very typical, being very large, sometimes measuring well over a metre along the outer edges of their usually crescentic form. Another characteristic feature of the horns is the presence of a series of 'ridges' across them. This animal is very fond of water and remains submerged in pools for a long time, with only the muzzle jutting out of the water.

As of now, the wild variety is not very large in number – somewhere around 2,000, but the domesticated ones are estimated to be well over 80 million. While the female is used as a source of milk, the male is used as a beast of burden. The hide, after treating, is used for making a variety of leather goods.

Of great significance to the present discussion, however, is the fact that the natural habitat of the water buffalo is South and South-East Asia. No

6. In these cases too, it needs to be pointed out that Griffith has wrongly translated the word *'mahiṣa'* as 'bull' instead of 'buffalo'. Anyway, what is more important is to note that the addition of the word *'mṛiga'* signifies the animal in its 'wild' state. The implication of the use of the term *'mṛiga'* along with *'hastin'*, which has been blown up out of all proportions by certain scholars, will be discussed in Chapter IV.

Fig. 3.4 *Mahiṣa* (*Bubalus bubalis*)

doubt at present one may come across this animal in certain other parts of the world, but it has been introduced there only lately. On the Indian subcontinent, on the other hand, it was domesticated as far back as 3,000 BC, if not earlier, as evidenced by its skeletal remains and representation in art at many sites of the Harappan Civilization. In fact, on a terracotta tablet from Harappa there is a very interesting scene which depicts this animal being offered as sacrifice to a deity that has been identified as Śiva (Fig. 3.5) – a practice still in vogue in parts of Himachal Pradesh. That the inhabitants of the Indian subcontinent were familiar with this animal from about the sixth-seventh millennium BC is indicated by the occurrence of the skeletal remains of the wild form of this animal in the aceramic neolithic levels (Period I) at Mehrgarh (Meadow 1984: 35).

Fig. 3.5 Harappa: Terracotta tablet. A person attempting to kill a buffalo (*mahiṣa*), perhaps offering sacrifice to the Śiva-like seated figure (right). Mature Harappan.

(iv) *Siṁha* (Lion, *Panthera leo* L.)

There are well over a dozen hymns in the *Ṛigveda* in which the *siṁha* has been mentioned. It has been recognized for its thunderous roar and regarded as the most dreaded animal. It moves fearlessly alike on plains and hills. Some of the verses referring to the *siṁha* are given below:

***RV* 1.64.8:**

siṁhā iva nānadati *prachetasaḥ piśā iva supiśo viśvavedasaḥ* /
kṣapo jinvantaḥ pṛiṣatībhirṛiṣṭibhiḥ samit sabādhaḥ śavasāhimanyavaḥ //

Exceeding wise they **roar like lions mightily**, they, all-possessing, are beauteous as antelopes;

Stirring the darkness with lances and spotted deer, combined as priests, with serpents' fury through their might.

There is yet another verse (***RV* 3.26.5)** which refers to the mighty roar of the lion:

agniśriyo maruto viśvakriṣṭaya ā tveṣamugramava īmahe vayam /
te svānino rudriyā varṣanirṇijaḥ ***siṁhā na heṣakratavaḥ*** *sudānavaḥ //*

The Maruts, Friends of men, are glorious as the fire: their mighty and resplendent succour we implore.

Those storming Sons of Rudra clothed in robes of rain, boon-givers of good gifts, **roar as the lions roar.**

In still another verse the thunder of the clouds, at the time of rainfall, is compared to the roar of the lion, which can be heard from a distance.

***RV* 5.83.3:**

rathīva kaśayāśvān abhikṣipannāvirdūtān kriṇute varṣyāṁ aha /
dūrāt siṁhasya stanathā udīrate *yat parjanyaḥ kriṇute varṣyam nabhaḥ //*

Like a car-driver whipping on his horses, he makes the messengers of rain spring forward.

Far off resounds the roaring of the lion, what time Parjanya fills the sky with rain-cloud.

There are references, such as the following, to the lion being a dreaded beast, chased by hunting bands and captured in snares:

***RV* 4.16.14:**

sūra upāke tanvaṁ dadhāno vi yat te chetyamritasya varpaḥ /
mrigo na hastī taviṣīmuṣāṇaḥ ***siṁho na bhīma*** *āyudhāni vibhrat //*

What time thou settest near the Sun thy body, thy form, Immortal One, is seen expanding:

Thou a wild elephant with might invested, **like a dread lion** as thou wieldest weapons.

***RV* 5.15.3:**

aṁhoyuvastanvastanvate vi vayo mahadduṣṭaram pūrvyāya /
sa sṁvato navajātastuturyāt ***siṁham na kruddhmabhitaḥ pariṣṭhuḥ*** *//*

Averting woe, they labour hard to bring him, the ancient, plenteous food as power resistless.

May he, born newly, conquer his assailants: **round him they stand as round an angry lion.**

***RV* 5.74.4:**

pauraṁ chiddhyudaprutaṁ paura paurāya jinvathaḥ /
yadīṁ gṛibhītatātaye ***siṁhamiva druhaspade*** *//*

Ye, Strengtheners, for Paura stir the filler swimming in the flood,
Advancing to be captured like a lion to the ambuscade.

***RV* 10.28.10:**

*suparṇa itthā nakhamā siṣā**yāvaruddhaḥ paripadaṁ nasiṁhaḥ*** /
niruddhaśchinmahiṣastarṣyāvān godhā tasmā ayathaṁ karṣadetat //

There hath the strong-winged eagle left his talon, **as a snared lion leaves the trap that caught him.**

Even the wild steer in his thirst is captured: the leather strap still holds his foot entangled.

The lion indeed is a majestic animal, drawing both awe and admiration (Fig. 3.6). Sometimes it is eulogized as 'the king of the jungle'. Well over a metre in height at the shoulder and about 2.5 m in length (excluding the long bushy tail), it weighs about 170 kg, which in some cases may go up to 250 kg. It has a broad head and the male is distinguished by the mane covering the neck and shoulders. While the mane may be black or brown in colour, the body shade is usually tawny yellow to reddish brown. The lioness is relatively smaller in size than the lion and also lacks the mane.

The lion preys on other animals like deer and zebras and even catches bigger ones such the buffalo and giraffe. It has thirty teeth, of which the four large canine ones are used to hold the prey and tear it away. Since there are no teeth suitable for grinding or chewing, the lion swallows in chunks the flesh of the hunted animal. Unless provoked, the lion usually does not attack humans. Its roar is so loud that it can be heard at a distance of about 7-8 km.

Fig. 3.6 *Siṁha* (*Panthera leo*). Natural habitat in India: Gir National Park, Gujarat.

Today the lion is met with in very limited regions of the globe. **It is found south of Sahara in Africa and in the Gir forest of India.** A couple of hundred years ago, it had been noted in other parts of India, such as Rajasthan and Haryana. There is evidence to suggest that in antiquity this animal inhabited parts of Mesopotamia and Iran as well.

(v) *Hastin / Vāraṇa* (*Elephas maximus* L. and *Loxodonta africana*)

In the *Ṛigveda* there are two words denoting the elephant, viz. *hastin* and *vāraṇa.* In both these cases the term *mṛiga* also occurs alongside, which indicates that it is the wild elephant that is meant in the contexts concerned. We shall deal later (Chapter IV, 58ff.) about the misinterpretation of the term *mṛiga* + *hastin* by certain scholars, whereby they have attempted to float a theory that the Aryans came from outside and thus on seeing the elephant having a trunk called it 'an animal with a hand'.

The verses in which the words *mṛiga* and *hastin/vāraṇa* occur in the *Ṛigveda* are given below:

***RV* 1.64.7:**

mahiṣāso māyinaśchitrabhānavo girayo na svatavaso raghuṣyadaḥ /

mṛigā iva hastinaḥ khādathā vanā *yadāruṇīṣu taviṣīrayugdhvam //*

Mighty, with wondrous power and marvellously bright, selfstrong like mountains, ye glide swiftly on your way.

Like the wild elephants ye eat the forests up when ye assume your strength among the bright red flames.

***RV* 4.16.14:**

sūra upāke tanvaṁ dadhāno vi yat te chetya mṛitasya varpaḥ /

mṛigo na hastī taviṣīmuṣāṇaḥ *siṁho na bhīma āyudhāni bibhrat //*

What time thou settest near the Sun thy body, thy form, Immortal One, is seen expanding:

Thou a wild elephant with might invested, like a dread lion as thou wieldest weapons.

***RV* 8.33.8:**

dānā mṛigo na vāraṇaḥ *purutrā charatham dadhe /*

nakiṣṭvā ni yamadā sute gamo mahāñścharasyojasā //

As a **wild elephant rushes on, this way and that way, mad with heat,**

None may compel thee, yet come hither to the draught: thou movest mighty in thy power.

***RV* 10.40.4:**

*yuvām **mṛigeva vāraṇā mṛigaṇyavo** doṣā vastorhaviṣā ni hvayāmahe /*
yuvam hotrāmṛituthā juhvate nareṣam janāya vahathaḥ śubhaspatī //

Even as the hunters follow **two wild elephants**, we with oblations call you down at morn and eve.

To folk who pay you offerings at appointed times, Chiefs, Lords of splendour, ye bring food to strengthen them.

We are sure, by now it would have been observed by the reader that in all cases **even Griffith translates the joint expression *mṛiga* + *hastin/vāraṇa* as 'wild elephant'**. In fact, there are many verses in the *Ṛigveda* wherein the word *mṛiga* by itself has been used to denote 'wild beast'. To these verses we would refer in Chapter IV when we discuss in detail the much exploited term *mṛiga* + *hastin* as an argument for advancing the theory that the Vedic Aryans were intruders from outside.

We may now discuss the physical features, habits and habitat of the elephant. This is the most massive of all the animals found on land. **There are two species, one inhabiting South and South-East Asia and the other Africa.** While the former may weigh up to 5,000 kg and measure up to 6.5 m in length and 3 m at the shoulder level (Fig. 3.7), the latter is a little more massive in all respects. Though similar in general look, they differ in details. Thus, for example, whereas the ears of the African elephant are very large, covering almost the entire shoulder, these are relatively smaller in the case of the Asian species (cf. Fig. 3.8). Again, the back of the African elephant has a dip in the middle, but that of the Asian is arched. Further, while the forehead of the African variety forms a smooth curve, that of the Asian elephant is distinguished by two humps. While in both the cases the legs are like massive pillars, there are differences in the number of toes: in respect of the hind legs, the Asian species has four toes, but the African only three. The long suspended trunk and tusks are features which set the elephant apart from the rest of the animals.

Wild elephants live in jungles and feed on tree-leaves and branches, shrubs, grasses, roots, bamboos, coconuts, etc. There are special methods of catching wild elephants and domesticating them. After domestication, they are used for

Fig. 3.7 Indian elephants (*Elephas maximus*).

riding and carrying heavy loads. In many areas they are used for uprooting trees from the nearby forests, which are then transported by them to the villages and towns. The tusk of the elephant is used for making a variety of ivory goods which may sometimes be sold at fantastic prices.

The elephant appears on the Mature Harappan seals, dating back to the middle of the third millennium BC (Fig. 3.9). Its skeletal remains have also been found at many sites of the Harappan Civilization, such as Mohenjo-daro, Harappa, Kalibangan, Lothal and Surkotada. At most of these sites have also been found objects made of ivory: combs, mirror-handles, hair-pins, antimony rods, etc. In fact, at some of the sites, for example Lothal and Monenjo-daro, there is evidence of ivory-working. That the inhabitants of the Indo-Pakistan subcontinent were familiar with this animal from a period much earlier than the Mature Harappan times is indicated by the occurrence of its faunal remains in Period I of Mehrgarh, ascribable to ca. seventh-sixth millennia BC (Meadow 1984: 35).

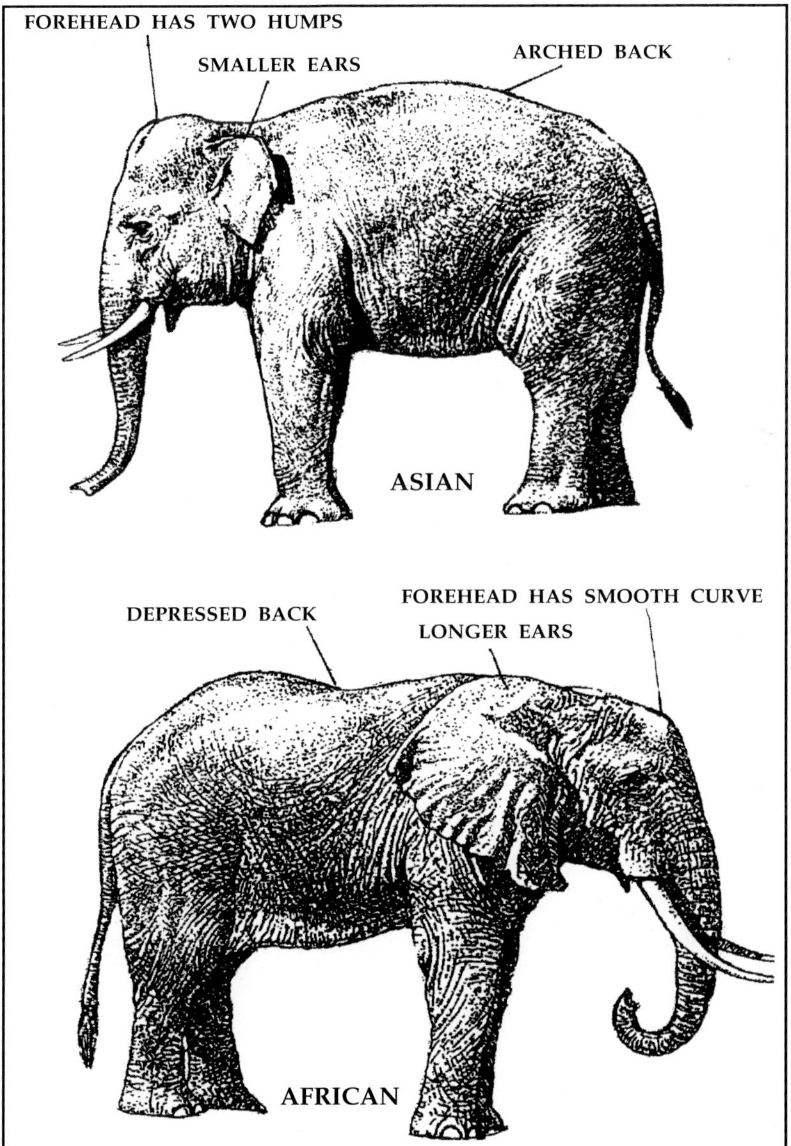

Fig. 3.8 Comparison of Asian and African elephants.

Fig. 3.9 Kalibangan: Seal and its impression, depicting the elephant.

B. BIRDS

(i) *Chakravāka (Anas casarca)*

Called the Brahmany duck in English and *Chakawā* in Hindi, the pair of this bird has been eulogized all through in Indian literature as an example par excellence of conjugal fidelity. The couple is supposed to be separated by the night but anxiously looks forward to the morning for a keenly awaited reunion. This aspect seems to be implied (though not explicitly stated) when, in a verse addressed to the two Aśvins, the Ṛigvedic poet requests them to come to him in a pair like the two *chakravāka*s. The relevant verse (***RV* 2.39.3**) is quoted below:

> *śṛiṅgeva naḥ prathamā gantamarvāk chhaphāviva jarbhurāṇā tarobhiḥ* /
>
> ***chakravākeva*** *prati vastorusrārvāñchā yātam rathyeva śakrā* //
>
> Like to a pair of horns come first to usward, like to a pair of hoofs with rapid motion;
>
> Come like two **Chakavās** in the grey of morning, come like two chariot wheels at dawn, ye Mighty.

I have not been able to procure a photograph of the *chakravāka* pair nor do I have details of their physical features, but **it seems that this bird is typically Indian**.

(ii) *Mayūra* (*Pavo cristatus* L.)

References to *mayūra* occur in ***RV* 3.45.1** and ***RV* 8.1.25**, while the female, *mayūrī*, is mentioned in ***RV* 1.191.14**. These are given below:

***RV* 3.45.1:**

ā mandrairindra haribhiryāhi ***mayūra****romibhiḥ* /

mātvā kechinni yamanvin na pāśinoti dhanveva tāṁ ihi //

Come hither, Indra, with Bay Steeds, joyous, with tails like **peacocks'** plumes.

Let no men check thy course as fowlers stay the bird: pass o'er them as o'er desert lands.

***RV* 8.1.25:**

ā tvā rathe hiraṇyaye harī ***mayūraśepyā*** /

śitipṛiṣṭhā vahatām madhvo andhaso vivakṣaṇasya pītaye //

Yoked to thy chariot wrought of gold, may thy two Bays with **peacock tails**,

Convey thee hither, Steeds with their white backs, to quaff sweet juice that makes us eloquent.

***RV* 1.191.14:**

Triḥ sapta mayūryaḥ *sapta svasāro agruvaḥ* /

tāste viṣam vi jabhrira udakam kumbhinīriva //

So have the **peahens three-times-seven**, so have the maiden Sisters Seven

Carried thy venom far away, as girls bear water in their jars.

Fig. 3.10 *Mayura* (*Pavo cristatus*).

Amongst the peafowls, **the Indian variety, *Pavo cristatus*, is the best known of the species** (Fig. 3.10). It is also found in Sri Lanka. These regions are its natural habitat and hence it is found there in forests and woodlands, besides, of course, being domesticated. However, it may also be seen in many other parts of the world, brought from elsewhere and usually housed in zoological parks. It is one of the most glamourous pheasants. The male, peacock, is distinguished by its wiry crest, greenish-blue neck, iridescent plumage and glittering train which is covered with multicoloured eye-like spots – light to deep blue, orange, etc. In the breeding season it is a sight to see the male when it erects its train in a fan-like manner. While the male may measure up to well over a metre, without taking the train into account, the female, peahen, is relatively shorter in size, has a smaller crest and no train. The peafowl feeds not only on grains, shoots, berries, etc. but is known to be equally fond of eating insects, even snakes. **Beyond India and Sri Lanka, the peafowl is found in other parts of South-East Asia, such as Myanmar, Malayasia, Indonesia.** There it is to be noted for its golden-green neck and breast.

On the Indian subcontinent there is ample evidence to establish that the peafowl was domesticated as far back as the third millennium BC, if not much earlier. It was a very favourite motif amidst the paintings on the Mature Harappan pottery (cf. Fig. 2.4), nay even still earlier as shown by examples from Kunal, an Early Harappa Culture site in the upper Sarasvati valley. There it was painted on an orange-red slip, in black outline and in-filled white colour, producing a kind of multi-chrome effect (cf. Fig. 2.3).

It is well known that in the third millennium BC India had trade contacts with the contemporary civilization of Mesopotamia, and I need not go here into the details thereof. However, what is important to note in the present context is the fact that Sargon of Akkad (ca. 2350 BC) mentions that in the harbour of his capital there were berthed boats from Dilmun, Maggan and Meluḥḥa; and scholars are inclined to identify Meluhha with the region of the Harappan Civilization. Further, there are documents which list up the items imported from the above-mentioned three regions. For example, in the Mesopotamian epic *Enki and the World Order*, referring to Meluhha, Enki states: 'May your bird be *haja*-bird, may it be heard in the royal palace'. Scholars have identified this bird as the peacock (cf. Ratnagar 1981: 69-70), though a few are not inclined to accept this identification. However, the peafowl seems to have been such an object of attraction in West Asia that it was depicted on a seal found in Bahrain, dated by the excavator to ca. 2000-2200 BC (Srivastava 1991: 27).

There is yet another region, besides South and South-East Asia where the peafowl is found. Known as *Afropavo congensis*, it occurs in **the Congo basin of Africa.** It is relatively smaller than the Asian variety, being at the most 70 cm in length. It has, on its head, a crest of black feathers, behind a tuft of white bristles.

CHAPTER IV

THE BEARING OF THE FOREGOING DATA ON 'THE ARYAN HOMELAND' DEBATE

In Chapter II we dealt with trees, plants and grasses that are mentioned in the *Ṛigveda*. These items, as already stated earlier, came to be referred to in the text not under a 'list of the flora', but only incidentally – for example, while dealing with the various parts of a chariot or vessels for certain rituals, and so on. However, these trees/plants give a fairly good idea of the kind of flora that existed on the land occupied by the Ṛigvedic people, which stretched from eastern Afghanistan on the west, via present-day Pakistan, to Indian Panjab, Rajasthan, Haryana and north-western Uttar Pradesh on the east. Alongside the Ṛigvedic verses that referred to a particular tree/plant was also given its global distribution, which highlighted the area of its occurrence and simultaneously, though inferentially, indicated the regions bereft of it.

In Chapter III the same exercise was done in respect of the fauna referred to in the *Ṛigveda*. Accompanying in each case of the animal/bird was also given its global distribution-pattern. This, again, indirectly brought to the fore areas which were not the natural habitat of the animal/bird concerned.

We shall now turn our attention to the possible implications of the foregoing data relating to the Ṛigvedic flora and fauna in terms of the Aryan 'homeland'. In doing so, I am afraid, quite a bit of what has already been stated in the above-mentioned two chapters will have to be recalled, in order to put side by side the data and the conclusions that may be drawn from them. However, efforts will be made to cut down the repetition to what is absolutely necessary. It is hoped that the reader will bear with us.

Aśvattha **(*Ficus religiosa* L.)**

As gleaned from ***RV* 1.135.8,** the wood of the ***Aśvattha*** was used for making vessels that would hold the Soma. It is also learnt from later texts that its sticks were used in kindling sacrificial fire: it is, thus, not unlikely that the same may have been the practice during the Ṛigvedic times. All this shows that the tree was well known to the Ṛigvedic people who used it for their everyday needs. The area of its occurrence is **eastern Afghanistan, Pakistan, India, Sri Lanka and Myanmar**.

Kiṁśuka / Parṇa **(*Butea monosperma* [Lamk.] Taub., Syn. *Butea frondosa* Roxb. / *Butea superba* Roxb.)**

According to ***RV* 10.85.20,** the ***Kiṁśuka*** wood, along with that of *Śalmali* (*Bombax ceiba* L.), was used for making chariots. This tree is found in **India, Pakistan, Bangladesh and Sri Lanka.**

Khadira **(*Acacia catechu* Willd.)**

Being fairly strong, the ***Khadira*** wood was also used for making chariots (***RV* 3.53.19**). It is **limited to India, Pakistan and Myanmar.**

Nyagrodha **(*Ficus benghalensis* L.)**

Although not mentioned specifically by its name, this tree seems to have been implied in verse **1.24.7** of the *Ṛigveda*. It is met with in **India, eastern Afghanistan, Pakistan and Sri Lanka.**

Vibhīdaka/Vibhītaka **(*Terminalia bellerica* Roxb.)**

It is gathered from ***RV* 7.86.6** that the nuts of this tree were used for dicing. The tree is at home in **Pakistan, India, Sri Lanka, Myanmar and maybe found as far east as Malaysia.**

Śalmali **(*Bombax ceiba* L., Syn. *Salmalia malabarica* [DC.] Schott.)**

The wood of this tree was also used in the construction of chariots (***RV* 10.85.20**). Its main region of occurrence is **Pakistan, India, Sri Lanka, Myanmar, but one may come across it even in Indonesia.**

Śiṁśapā **(*Dalbergia sisso* Roxb.)**

Because of its sturdiness, the wood of this tree was also used in chariots (***RV* 3.53.19**). Its provenance is **India, Pakistan and Afghanistan.** The small region of south-eastern Iran, where too one may come across it, may, for all intent and purposes, be regarded as an extension from Afghanistan.

From the foregoing it would be clear that most of these trees belong to the Indian subcontinent, with a spill over to Afghanistan on the one hand and (sometimes) to what is known as South-East Asia on the other.

Turning our attention to plants and grasses, the position may be summarized as follows:

***Urvāruka* (*Cucumis melo* L. var. *utilissimus* Roxb. Duthie et Fuller).** In the Ṛigvedic hymn concerned (**7.59.12**), this plant is mentioned in a philosophic context. The devotee's prayer is that he/she may be freed from the bondage of death just as the ***urvāruka* fruit,** as soon as it is ripe, is released from the plant itself; and he/she may be granted immortality. **Growing wild in India, Pakistan and tropical Africa, it is cultivated, besides these countries, in Afghanistan, Iran, Malaysia and even north Australia.**

Darbha **(*Imperata cylindrica* [L.] P. Beauv. Syn. *Saccharum cylindricum* [L.] Lamk.)**

Since the word ***darbha*** occurs in the Ṛigvedic verse **(*RV* 1.191.3)** along with other grasses, it is evident that it belonged to that category. However, its exact Latin equivalent is not very certain. At the same time, the context indicates that these grasses **grew luxuriantly in wilderness, sheltering poisonous reptiles. A hot and humid climatic zone is thus indicated.**

Pākadūrvā **(*Cynodon dactylon* [L.] Pers., Syn. *Panicum dactylon* L.)**

Along with two other grasses, viz. *kiyāmbu* and *vyalkaśā* (which unfortunately we have not been able to identify), the ***Pākadūrvā*** is stated in ***RV* 10.16.13** to grow in areas where the dead body is cremated. Its provenance is **Pakistan, India, Sri Lanka and Myanmar.**

Muñja **(*Saccharum bengalense* Retz., Syn. *S. munja* Roxb.)**

It is referred to in the *Ṛigveda* in two different contexts. In ***RV* 1.191.3** it is stated that venomous creatures inhabit the area where this grass grows, along with a few others, such as *sairya* and *vīraṇa* which, we regret we have not been able to identify. ***RV* 1.161.8** mentions the use of ***Muñja*** in the filtering of the Soma. It grows in **Afghanistan, Pakistan, India, Bangladesh and even in some parts of China**.

Śīpāla **(*Blyxa octandra* Planch. Syn. *B. Roxburghii* Rich.)**

Referred to in ***RV* 10.68.5**, this plant grows in still water, such as in tanks or small lakes. It grows in the **tropical regions of Asia and Australia.**

Soma **(Species of *Sarcostemma* L. ? / *Ephedra* L. ?)**

This plant is very much eulogized in the *Ṛigveda*, so much so that a whole *maṇḍala* (viz. 9) is devoted to it. Its juice was offered to gods in rituals. Although it has not been possible to identify this plant with certainty, from all accounts it seems to have grown in the **hills between the north-western part of the Indian subcontinent and Afghanistan.**

A look at the distribution-pattern of the foregoing plants and grasses indicates that these too, like the above-mentioned trees, are at home in tropical and sub-tropical climates.

At this point it seems necessary to examine the statement of Possehl (1996: 65) to which a preliminary reference had already been made in the Introduction (above, pp. 1-2). To recall, he says: 'One thing seems certain; the **speakers of Vedic Sanskrit** are the earliest *well-documented* (italicized by Possehl himself) speakers of an Indo-European language in the subcontinent, and they **came from elsewhere.**' One of the arguments he puts forward in support of his theory is a reference to 'Indo-European words for trees which are species such as birch, Scotch pine, linden, alder and oak. These are plants from a temperate environment and the fact that their names are shared among the early languages of the family suggests a homeland in this environment (Friedrich 1970, specially pages 152-58).'

On the one hand, as discussed in the preceding pages, **all of the Ṛigvedic trees belong to a tropical environment and, on the other, none of the five 'temperate-environment' trees mentioned by Possehl, viz. birch, Scotch pine, linden, alder and oak, finds a mention in the *Ṛigveda*. *The Ṛigvedic flora does not even reflect a memory of such trees.* Then how, on the basis the aforementioned five trees, can the theory that the Ṛigvedic Aryans 'came from elsewhere' be sustained ?**

And here it may be well worth while to add that even if trees such as the birch (*bhurja*), pine (*chīra*) and cedar (*devadāru*), which belong to a temperate climate, were mentioned in the *Ṛigveda,* that would have been no basis to argue that the Ṛigvedic people came from outside. These trees are met with also in the mountainous regions of north-western parts of the Indian subcontinent – which was the home of the Ṛigvedic people, as clearly revealed by the famous *Nadi-stuti Sukta* of the *Ṛigveda* (***RV* 10.75. 5-6)**: it defines their habitat as from the Gaṅgā -Yamunā on the east to the Indus, along with its north-western tributaries, on the west. In fact, ***RV* 10.121.4 even refers to snow-clad mountains, if not actually to the Himalayas themselves.**

> *Yasyeme* ***himvanto*** *mahitvā yasya samudram rasayā sahāhuḥ /*
>
> *yasyemāḥ pradiśo yasya bāhū kasmai devāya haviṣā vidhema //*
>
> His, through his might, are these **snow-covered mountains**, and men call sea and Rasā his possession:
>
> His arms are these, his are these heavenly regions. What God shall we adore with our oblation ?

And if we care to have a look at the archaeological evidence, it also shows that as early as the third millennium BC the Indians were exploiting the Himalayan flora. For wasn't the shroud/lid of a coffin in one of the graves at Harappa made of deodar (*Cedrus deodara*) which grows in the Himalayan region (Choudhury and Ghosh 1951) ?

We may now turn our attention to the **animals** in the *Ṛigveda.*

Uṣṭra (Camel, *Camelus dromedarius* and *Camelus bactrianus*)

There are about **half-a-dozen verses in the *Ṛigveda*** in which the word ***uṣṭra*** occurs. One of these **(*RV* 8.5.37)** refers to the gift of a hundred camels, while in another **(*RV* 8.6.48)** it is stated that these were yoked in fours. There are two species of the camel. One, having a single hump and known as ***Camelus dromedarius***, is found **from India on the east to Arabia on the west**, whereas the other species, with two humps and called ***Camelus bactrianus***, inhabits **Afghanistan, Mongolia and even parts of China.** In the *Ṛigveda*, however, there is no specific mention about the number of humps.

Gaura (*Bos gaurus*)

Both *Gaura,* the male, and *Gaurī,* the female, have been referred to in the *Ṛigveda*: the former in about half-a-dozen verses (e.g. ***RV* 4.21.8; 7.69.6**) and the latter in at least three (e.g. ***RV* 1.84.10**). Though Griffith has mistranslated the word, sometimes as 'stag' and at others as 'buffalo', the correct identification is ***Bos gaurus.*** Heavily built and characterized by a prominent ridge above the shoulder, this animal is found in **India, Nepal, Myanmar, Thailand, Malaysia and even parts of China.**

Mahiṣa (Water buffalo, *Bubalus arnee/Bubalus bubalis* L.)

In the *Ṛigveda,* the ***mahiṣa*** (domesticated = ***Bubalus bubalis***) is mentioned in a number of verses, such as ***RV* 5.29.7; 9.87.7**. When the wild form (***Bubalus arnee***) is intended, the word ***mahiṣa*** is accompanied by the term ***mṛiga,*** for example in ***RV* 9.92.6; 10.123.4**. Though the buffalo may now be seen in many parts of the world (having been introduced there recently), **its natural habitat is South and South-East Asia.** On the Indian subcontinent, it was depicted in the Harappan art of the 3rd millennium BC, while its skeletal remains have been found in the Neolithic levels at Mehrgarh, datable to the 6th-7th millennia BC.

Siṁha (Lion, *Panthera leo* L.)

Regarded as the most dreaded animal, with a thunderous roar, the ***siṁha*** has been referred in well over a dozen verses in the ***Ṛigveda,*** e.g. **1.64.8; 4.16.14, etc.** There is a reference even to its being caught in a trap (***RV* 10.28.10**). **As of now, in the wild state the lion is met with in the Gir Forest of India and south of Sahara in Africa.**

Hastin/Vāraṇa (*Elephas maximus* L. and *Loxodonta africana*)

Both the above-mentioned words, viz. ***Hastin*** and ***Vāraṇa,*** occur in the ***Ṛigveda*** to denote the elephant. In order to indicate that in the contexts concerned it is the wild elephant that is meant, the word ***mṛiga*** was used in addition (***RV* 1.64.7; 8.33.8**). Size-wise, the elephant is the biggest animal found on land. There are two sub-species: one, known as ***Elephas maximus,* inhabiting South and South-East Asia** and the other, called ***Loxodonta africana,* found in Africa.**

And now to the **birds**.

Chakravāka (*Anas casarca*)

In ***RV*** **2.39.3,** the devotee prays to the two Aśvins to come to him like the pair of ***Chakravāka***s that has all along been referred to Indian literature as a symbol of great mutual love and fidelity. It seems to be a typical **Indian bird.**

Mayūra (Pavo Cristatus L.)

Both the ***Mayūra*** (the male, peacock) and ***Mayūrī*** (the female, peahen) have been referred to in the *Ṛigveda*: the former in ***RV*** **3.45.1 and 8.1.25** and the latter in ***RV*** **1.191.14.** There are two sub-species of this bird. One, known as ***Pavo cristatus,*** is met with in **India, Sri Lanka, Myanmar, Malaysia and Indonesia.** The other, called ***Afropavo congensis,*** inhabits the **Congo basin of Africa.**

Before winding up this summary of the Ṛigvedic fauna, it seems necessary to examine the views of certain scholars regarding the term *'mṛiga + hastin'* used in the *Ṛigveda,* for example Roth in *St. Petersburg Dictionary* and *Nirukta, Erlauterungen,* 79. The term has been taken to mean 'an animal with a hand', which epithet has been interpreted to suggest that the elephant, whose trunk looks like a hand, was new to the Vedic Aryans. As a further step in this interpretation, it has been argued that this term was coined by the Aryans when they entered India from the cold climes of Central Asia and saw a new kind of animal whose trunk looked like a hand. Thus, the use of this term began to be cited as an additional proof of the long-cherished theory that the Aryans came from Central Asia.

We may now examine how far this kind of conclusion is valid. The word *'hastin'* accompanied by the word *'mṛiga'* occurs in two passages of the *Ṛigveda,* viz. **1.64.7** and **4.16.14,** which have already been referred to earlier (p. 44) but are quoted once again, along with Griffith's translation, for the sake of ready reference.

> *mahiṣāso māyinaśchitrabhānavo girayo na svatavaso raghuṣyadaḥ* /
>
> ***mṛigā iva hastinaḥ khādathā vanā*** *yadāruṇīṣu tavaṣīrayugdhvam* // **1.64.7**
>
> Mighty, with wondrous power and marvellously bright, self-strong like mountains, ye glide swiftly on your way.

Fig. 4.1 *Siṁha* (Lion) is an example of one who eats directly with the mouth (*mukhādāna*).

Like the *wild* elephants ye eat the forests up when ye assume your strength among the bright red flames.

sūra upāke tanvaṁ dadhāno vi yat te chetyamṛitasya varpaḥ /

mṛigo na hastī tavaṣīmuṣāṇaḥ *siṁho na bhīma āyudhāni bibhrat //* **4.16.14**

What time thou settest near the Sun thy body, thy form, Immortal One, is seen expanding:

Thou **a *wild* elephant with might invested,** like a dread lion as thou wieldest weapons.

Fig. 4.2 In contrast to the lion, the elephant (=Sanskrit *Hastin,* meaning that which has a *hasta,* i.e. a hand) eats with its 'hand', i.e. the trunk (*hastādāna*).

In both these verses the additional term ***'mṛiga'*** has been used to denote **'wild'**. In **1.64.7,** the simile is to the 'wild' elephant who eats away the forests – even now a common sight in the sub-Himalayan forests. In **4.16.14**, the 'wild elephant' has been likened to a 'dreaded lion', obviously because of their commonality in terms of extraordinary strength. It is thus abundantly clear that the supposed newcomers (the Aryans) did not coin the word *'mṛiga hastin'* to denote an animal previously unknown to them by calling it 'an animal with a hand'. The word *'mṛiga'* comes from the root *'mṛig'* which means 'to hunt', and hence the sense of 'wild'.

In this context, it may not be out of place to recall that another word for the elephant, viz. *'vāraṇa'*, was qualified with the adjective *'mṛiga'* when the animal's 'wildness' was intended to be conveyed. See, for example, **RV 8.33.8**:

dānā mṛigo na vāraṇaḥ purutrā charatham dadhe/

nakiṣṭvā ni yamadā sute gamo mahāṁścharasyojasā //

As a *wild* elephant rushes on, this way and that way, mad with heat,

None may compel thee, yet come hither to the draught: thou movest mighty in thy power.

There are quite a few verses in the *Ṛigveda* wherein the word *mṛiga* has been used by itself to denote a wild beast; and here we quote some of these.

***RV* 1.191.4:**

ni gāvo goṣṭhe asadan ***ni mṛigāso avikṣata*** */*

ni ketavo janānām nyadṛiṣṭā alipsata //

The cows had settled in their stalls, **the beast of prey had sought their lairs,**

Extinguished were the lights of men, when things unseen infected me.

***RV* 8.1.20:**

mā tvā somasya galdayā sadā yāchannaham girā /

bhūrṇim mṛigam na savaneṣu *chukrudham ka īśānam na yāchiṣat //*

Let me not, still beseeching thee with earnest song at Soma rites,

Anger thee like **some wild beast.** Who would not beseech him who hath power to grant his prayer.

That the term *'mṛiga'* was indeed applied to convey the 'wild' character of the animal concerned becomes once again clear when an expression like *'puruṣa mṛiga'* was used to denote an 'ape' in the *Yajurveda* (*Maitrāyaṇī Saṁhitā* 3.14.16), since the literal meaning would be 'man wild beast'. This was so done since the ape resembled a human being but was 'wild'.

All told, therefore, it would be wrong to argue that the occurrence of an expression like *'mṛiga hastin'* in the *Ṛigveda* upholds the assumption that the

Aryans came from outside and on seeing the elephant were so much excited that they cried out: 'Look, here is an animal with a hand'.

The Vedic people seem to have had a valid reason to call the elephant '*hastin*'. It evidently was because this animal was different from the other animals – in having **a trunk which functioned as a hand.** The *Maitrāyaṇī Saṁhitā* (4.5.7) and the *Taittirīya Samhitā* (6.4.5.7) give an interesting **division of animals into two broad categories.** In one are those that hold objects (or eat) directly with the mouth, the term used being ***mukhādāna.*** Belonging to this category, for example, are the cattle, lion (Fig. 4.1), etc. In the other category (***hastādāna***) are the ones that do so with the aid of their hands, such as the ape (*markaṭa*), man (*puruṣa*) and the elephant (*hastin*). Hence the validity of the term '*hastin*' (Fig. 4.2). Clearly, no other 'strings' need to be 'attached' to the word. Given below is the relevant extract from the ***Taittirīya Saṁhitā*** **(6.4.5.7)** edited by S.D. Satvalekar, Pardi, 1957:

> *brahmavādino vadanti kasmāt satyāt trayaḥ paśūnām hastādānā iti yat trirūpāṁśuṁ* ***hastena vigṛihṇāti*** *tasmāt trayaḥ paśūnām hastādānāḥ puruṣo* ***hasti*** *markaṭaḥ* /

The distribution-pattern of the Ṛigvedic flora and fauna, as discussed above, shows beyond any shadow of doubt that their *natural habitat is the tropical zone and not the cold one.* Indeed, there is no mention whatsoever of any typically cold-climate tree/plant or animal/bird in the *Ṛigveda. Nor is in any way even their memory reflected. Thus, both the positive and negative kinds of evidences show that the Ṛigvedic Aryans had nothing to do with a cold-climate zone.*

CHAPTER V

THE HOMELAND OF INDO-EUROPEAN LANGUAGES AND CULTURE: SOME THOUGHTS*

Delivering the Third Anniversary Discourse at the meeting of the Royal Asiatic Society of Bengal in the year 1786, Sir William Jones made a very significant statement, as follows (Jones 1788):

> The Sanskrit language, whatever may be its antiquity, is of wonderful structure; more perfect than the Greek, more copious than the Latin, and more exquisitely refined than either, yet bearing to both of them a stronger affinity both in the roots of verbs and in the forms of grammar, than could have been produced by accident; so strong, indeed, that no philologer [sic] could examine them all three, without believing them to **have sprung from some common source**, which perhaps no longer exists....

This discovery of a similarity amongst the above-noted three languages – Sanskrit, spoken in India on the east and Greek and Latin in southern Europe on the west – led immediately to the formulation of a theory that there must have existed a 'Proto-Indo-European language' which was parental to all these languages.

A formulation of the branches and sub-branches of this hypothetical Proto-Indo-European language has been made out by some scholars as follows:

---

* Based on a paper presented by the author at an International Conference on 'The Homeland of Indo-European Languages and Culture', held at New Delhi on January 7-9, 2002, under the auspices of the Indian Council of Historical Research.

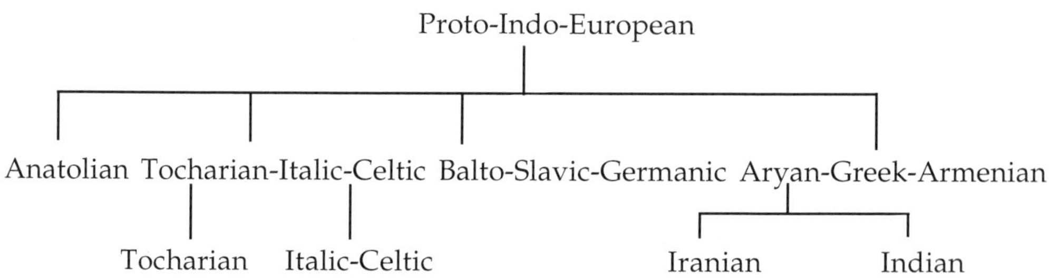

The concept of a Proto-Indo-European language led to another concept, namely that of a 'Proto-Indo-European race'. As a sequel to this assumption, began the hunt for the 'original home' (*Urheitmat*) of this 'race'.

Initially, some scholars opined that India, being the home of the earliest extant literature (viz. the Vedas) of the Indo-European group of languages, must have been the original homeland of these languages and hence of the hypothesized Indo-European race. However, soon the canvas got enlarged and, leaving India aside, the scenario was taken all the way to Europe. Not surprisingly, almost every part thereof was declared to be the homeland: Scandinavia, Finland, south-west Russia, the Baltic area, Germany, Lithuania, Hungary, the Danube valley and so on. In fact, the rat race was so much that no part of Europe was left out. In a very sarcastic comment on this race, Jean-Paul Demoule (1980: 120) averred: 'We have seen that one primarily places the IE's [Indo-Europeans] in the north if one is German .... in the east if one is Russian, and in the middle if, being Italian or Spanish, one has no chance of competing for the privilege.'

However, in the course of time this Euro-centric approach began its climb-down and many new regions were upgraded. As of now, the more important claims to the Indo-European homeland pertain, in their geographical locations from west to east, to (Fig. 5.1) : (i) the Anatolian region of western Asia; (ii) the Black Sea-Caspian Sea belt; (iii) the steppes of southern Russia; and (iv) Sogdiana in south-central Asia. We shall deal with these claims first, after which the possibility of north-west South Asia having been the homeland will also be considered (E).

The first question that has to be addressed while looking for the original home of these Proto-Indo-Europeans is at what cultural level were they: hunter-

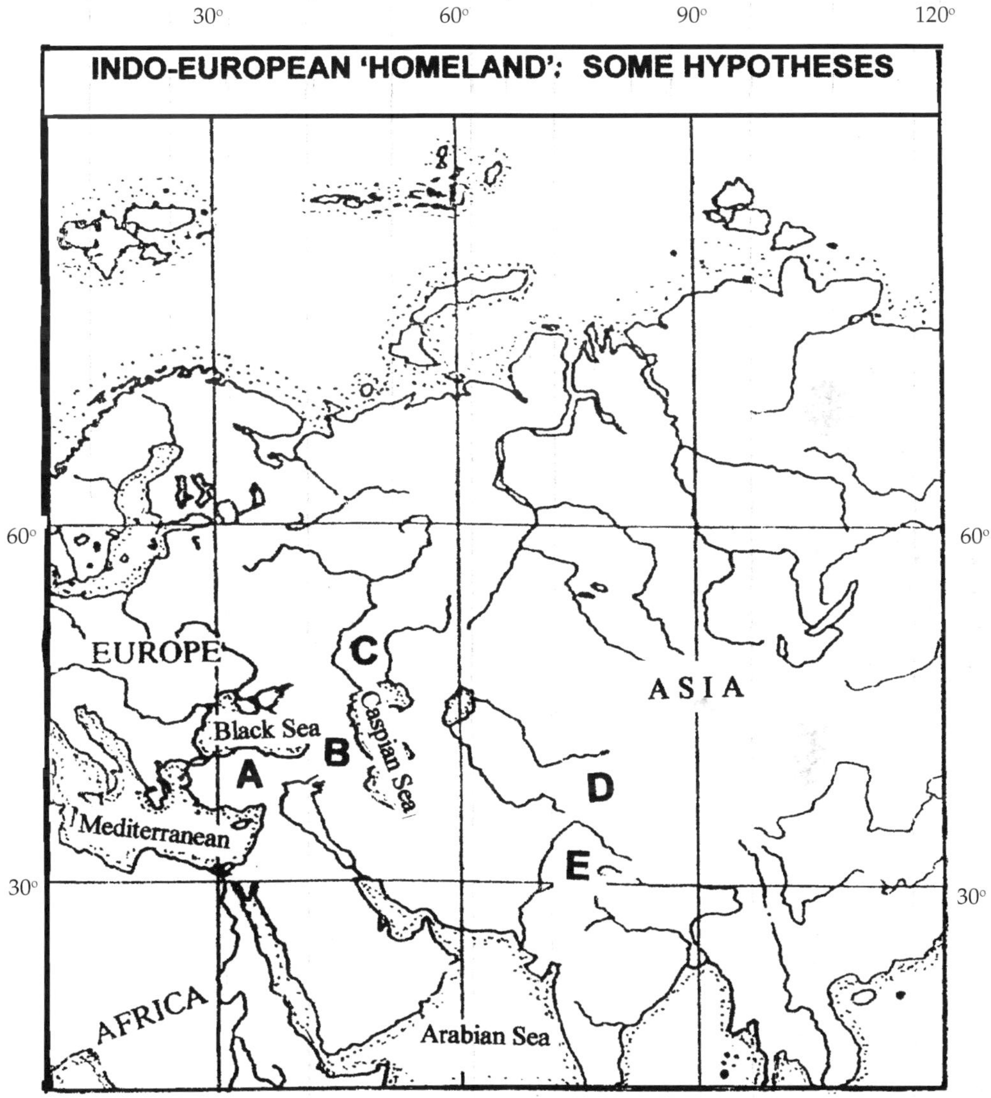

A = Anatolia; B = Caucasus; C = Kurgan; D = Sogdiana; E = NW South Asia

Fig. 5.1 Indo-European Homeland: Some hypotheses.

gatherers, pastoral nomads or sedentary agriculturists? An attempt in this direction was made by trying to find out the common words pertaining to agriculture, animals, plants, climate, etc. in the various Indo-European languages and thereby to determine the techno-cultural level and the geophysical surroundings of these hypothetical Proto-Indo-Europeans. Since the reconstructed terms in many cases are faulty, no consensus on this issue could be arrived at. Thus, some scholars would like to visualize these Proto-Indo-Europeans as nomads roaming about the steppes in cold climate, whereas others would regard them as sedentary agriculturists, living in an area with a moderate climate and endowed with wild cereal plants and animals that could easily be domesticated.

(i) The Anatolian Homeland

The noted proponent of the Anatolian homeland is Colin Renfrew who believes (1987) that the Proto-Indo-Europeans are first identifiable in Anatolia where they practised agriculture around 7000 BC. And it was from here that one of their groups moved westward to Europe and another group, moving eastward, via the region south of the Caucasus mountains and the Caspian Sea, reached Iran from where it must have subsequently entered Afghanistan and India. In an alternative scenario, Renfrew thinks that the Proto-Indo-Europeans split up **after** entering Europe and then the eastern branch went to south-central Asia, via north of the Black and Caspian Seas, whence it moved on to north-eastern Iran, Afghanistan and India. The Anatolian hypothesis, however, falters on at least two major counts. In the first place, if the Europeans, on the one hand, and the Indo-Iranians, on the other, had once lived together as agriculturists in Anatolia, they ought to have a common vocabulary for agricultural items, which unfortunately is **not** the case. Secondly, the language of Anatolia, on which this commonness has been perceived, was a 'minority' language, probably of the elites, whereas the basal language was non-Indo-European. This scenario is hardly tenable with the concept of the Indo-Europeans having been the original inhabitants of this area. On the Anatolian agricultural setting vis-à-vis the Indo-Europeans, there comes a bitter comment, from across the Atlantic, by Lamberg-Karlovsky (1988: 2) who snaps: ".... [the whole issue has been] simplified by Professor Renfrew to the **ludicrous formula** 7000 B.C.E. Anatolia = farming = Indo-Europeans."

(ii) The Caucasus Region

Shifting the focus to the east of Anatolia, Gamkrelidze and Ivanov (1995) place the homeland in the region between the Black and Caspian Seas. Their thesis is based primarily on linguistic palaeontology, applying which they aver that the homeland was a mountainous terrain, replete with lakes and fast-flowing rivers, which their proposed region, they argue, has in ample measure. In an attempt to add further weight to their thesis, they argue that since the reconstructed Indo-European language has a number of Semite loan-words, the homeland could not have been far away from the Semitic world. However, the thesis of there being a large number of Semitic loan-words in the Indo-European language has been shown by many a scholar as a misplaced belief. Besides, there is another linguistic incongruity in the thesis, viz. that the language spoken in the area between the Black Sea and the Caspian Sea is replete with a strong element of non-Indo-European vocabulary, suggesting that there was a sizable substratum of non-Indo-Europeans. Consequently, Indo-Europeans **could not** have been the original inhabitants of the area.

(iii) The 'Kurgan' Homeland

In yet another scenario, the venue is shifted to the steppes lying to the north of the Black and Caspian Seas. Over here archaeological remains of a culture typified by burial barrows (called *'kurgan'* in the Slavic language) have been met with. The chief proponent of this thesis is Maria Gimbutas (1966 and 1997). According to her, the reconstructed linguistic evidence suggests that the Indo-Europeans were horse-riding warriors who used thrusting weapons and could easily overrun other areas, and that they did do so insofar as central Europe is concerned, around the fourth-fifth millennia BC. However, on the techno-cultural level, the Kurgan people were essentially at a pastoral stage. Hence, discounting this equation, Renfrew (1999: 268) holds that on the European scene mounted warriors appear only as late as the turn of the 2nd-1st millennia BC and these could in no case have been Gimbutas's Kurgan warriors predating the facts by some 3000 years. On the linguistic turf, there comes a severe attack by Kathrin Krell (1998) who finds a great incongruity between the terms found in the reconstructed Indo-European language and the cultural level met with in the kurgans. For example, Krell holds that the Indo-Europeans had reached an agricultural level whereas the Kurgan people were just at a pastoral stage.

There are others, like Mallory and Schmitt, who are equally critical of Gimbutas's hypothesis.

(iv) The Sogdiana Homeland

Descending to the south-east, another homeland has been suggested, 'somewhere in the vicinity of ancient Bactria-Sogdiana', by Johanna Nichols (1997 a and b). From this homeland, Nichols holds, there was a spread of the Proto-Indo-European language to the area surrounding the Aral Sea and on to the Caspian. From there a two-fold spread has been envisaged: a major one to the areas lying to the north of the Caspian and Black Seas and thence to Europe, and a comparatively minor one along the southern side of these seas, also reaching Europe via Anatolia, from the southern end. However, a more noteworthy point in Nichols's schema is that there was only a language-spread and not a migration of people. While such a stand absolves the proponent from producing proof in terms of the material culture as well, one has yet to be fully convinced of the hypothesis that a language can keep on spreading from area to area, without involving the language-carriers, viz. the people themselves. However, Nichols's model, proposed only recently, has yet to be fully evaluated by linguists.

In the preceding pages we have seen the 'kurganization' (i.e. burial under barrows) of three of the four theses, viz. those relating to Anatolia, Caucasus and the Russian steppes. Also, as already mentioned, the Sogdiana one has yet to stand some rigorous tests. We have also noted that in the case of all the three aforementioned theses, for each proponent there are at least half-a-dozen opponents. A very interesting point that emerges from these controversies is that the dispute is not merely between archaeologists on the one hand and linguists on the other, who flaunt their own discipline as being superior to that of the other, but amongst archaeologists themselves and similarly amongst linguists, indicating that not even two practitioners of the same discipline see eye to eye. Completely disillusioned with such a scenario, Mallory (1989: 143) rightly observed: "One does not ask 'where is the Indo-European homeland?' but rather **'where do they put it now?'**" (Emphasis mine.)

(v) North-West South Asia

On my part, however, I would not like the search to be given up and would indicate **the region to go to now.** Let us move a full circle and try out the Indian homeland thesis which was proposed long back but could not hold the ground. I am aware that there would be an instantaneous uproar at this proposal, but why be allergic to the very idea itself? The reason for a fresh examination of this proposal is that in those early days there was a total absence of any archaeological data from north-west South Asia, which we now have in abundance on account of excavations at a large number of sites. **To be more precise, my proposal includes the region which starts from Afghanistan on the west, includes present-day Pakistan, Indian Punjab, Haryana, Rajasthan and ends up in north-western Uttar Pradesh on the east. Also, it needs to clarified that it *does not include* other parts of India wherein there existed other contemporary ethno-cultural entities.**

The evidence from the excavations at Mehrgarh in Pakistan (Jarrige *et al.* 1995) has demonstrated that the north-western part of the Indian subcontinent had reached a neolithic, i.e. settled agricultural, stage by the 7th millennium BC. Here it may also be emphasized that the Mehrgarh neolithic complex stands in marked contrast to that of Western Asia. For example, whereas in the West Asian neolithic there is the dominance of sheep and goat amongst the domesticated animals, in the Mehrgarh context the cattle dominated over other animals. Further, no less noteworthy is the presence of the humped bull (zebu, *Bos indicus*) amidst the cattle remains of Mehrgarh, right from Period I (Meadow 1984: 37). And in Period II, even a terracotta figurine of that animal was found (Meadow *ibid*: 36). This is a species that is typical of the Indian subcontinent and has little to do with West Asia. Likewise, the water buffalo, which is so typical of the Indian subcontinent and Southeast Asia and was rare in West Asia, has been found at Mehrgarh, first in its wild variety (*Bubalus arnee)* in the pre-ceramic neolithic levels (IA) and later in the domesticated form (*Bubalus bubalis*). In terms of the cultivated cereals, it is wheat that dominated in the West Asian neolithic, but in the Indian context, it was barley and not wheat. Thus, the Mehrgarh neolithic has its own identity, having no generic relationship with its West Asian counterpart. In other words, the Mehrgarh people were the **'the sons of the soil'**.

Further, there is a continuous story from the succeeding chalcolithic levels onwards, taking us through various evolutionary stages to the Early Harappan from which there emerged the Harappan Civilization itself, around the middle of the third millennium BC. Again, after a thorough study of the human skeletal remains, Hemphill and his colleagues (1991) have shown that there was a human biological continuity right from 4500 BC to 800 BC. A question may now be posed: 'What language did these chalcolithic people speak?' Though, in spite of so many tall claims, the Harappan script still remains undeciphered (cf. Lal 1970, 1974 and 1983), yet we have another way of tackling the issue.

In the *Ṛigveda*, the Sarasvati has been stated to be a mighty river, flowing from the mountains to the sea. For example, ***RV* 7.95.2** states:

> ***ekāchetat Sarasvatī nadīnām śuchir yatī giribhya ā samudrāt /***
>
> *rāyaśchetantī bhuvanasya bhūrerghṛitam payo duduhe Nāhuṣāya //*
>
> Pure **in her course from mountains to the ocean,** alone of streams **Sarasvatī** hath listened.
>
> Thinking of wealth and the great world of creatures, she poured for Nahuṣa her milk and fatness.

***RV* 6.61.2** mentions how powerful this river was:

> ***iyam śuṣmebhir bisakhā ivārujat sānu girīṇām taviṣebhirūrmibhiḥ /***
>
> *Pārāvataghnīmavase suvṛiktibhiḥ* ***Saratsvatīmā*** *vivāsema dhītibhiḥ //*
>
> **She with her might, like one who digs for lotus-stems, hath burst with her strong waves the ridges of the hills.**
>
> Let us invite with songs and holy hymns for help **Sarasvatī** who slayeth the Pārāvatas.

By the time of the *Panchviṁsa Brāhmaṇa* (XXV.10.16) it dried up. 'At a distance of a journey of forty days on horseback from the spot where the Sarasvatī is lost (in the sands of the desert), (is situated) Plakṣa Prāsravaṇa.' (Caland's translation, 1931, reprint 1982, p. 636.)

When did this drying up of the Sarasvati take place? The answer is provided by the evidence from the excavations at Kalibangan which stood on the bank of the Sarasvatī, now going by the name of the Ghaggar.

However, before taking up the Kalibangan evidence, it seems necessary to fix up the identity of the Ṛigvedic Sarasvatī, since many misconceptions have been floated by certain scholars about its identity. Thus, according to some, it was the Helmand of Afghanistan or even the Indus itself. Says R.S. Sharma (1999: 35): 'The **fundamentalists** want to establish the superiority of the Sarasvati over the Indus because of **communal** considerations.' What a baseless charge! Anyway, he then asserts that 'The earliest Sarasvati is considered identical with the **Helmand of Afghanistan.'**

Griffith, in the course of his translation of the *Ṛigveda* (1995 reprint, Vol. I, p. 631, footnote) observes: '... it seems probable that the Sarasvatī is another name of the Sindhu.'

Depending solely on the translation by Griffith of ***RV* 10.64.9**, the Habibs (1991-92: 76) assert. 'But it seems likely .... that Sarasvatī/Harakhvati was the name for the Helmand, above its junction with Arghandab. Otherwise, it is difficult to understand how Sarasvatī could be put **between Sindhu and Sarayu,** i.e. the Indus and Hariud, as the major intervening river in RV 10.64.9.' The Habibs forget that Griffith was constrained in his translation of the passage by **the requirements of metrical versification**, and the fact of the matter is that **the Ṛigvedic text *does not* put the Sarasvatī *between* the other two rivers.** It runs as follows: ***Sarasvatī Sarayuḥ Sindhurūrmibhir maho mahīravasā yantu vakṣaṇīḥ.*** Thus, any conclusion based on Griffith's aforesaid translation is basically wrong.

Further, since the foregoing passage clearly mentions the Sarasvatī and Sindhu by separate names, no extra wisdom is needed to establish that these are two separate rivers. And this is not the only verse where these two rivers are mentioned separately. The famous *Nadī–stuti* hymn (***RV* 10.75**) does it in a big way.

In fact, this very hymn provides the most unimpeachable evidence in respect of the geographical placement of the Ṛigvedic Sarasvatī. Since it is of utmost importance, we quote here the relevant two verses from it :

imam me ***Gaṅge Yamune Sarasvati Śutudri*** *stomam sachatā Paruṣṇyā /*

Asiknyā Marudvṛidhe Vitastayā Ārjikīye śṛiṇuhyā Suṣomayā //5//

Tṛiṣṭāmayā prathamam yātave sajūḥ Susartvā Rasayā Śvetyā tyā /

Tvam ***Sindho*** *Kubhayā Gomatīm Krumum Mehatnvā saratham yābhirīyase //6//*

O Gaṅgā, Yamunā, Sarasvatī, Śutudrī (Sutlej) and Paruṣṇī (Ravi), O Marudvṛidhā with Asiknī (Chenab), O Ārjikīyā with Vitastā (Jhelum) and Suṣomā (Sohan), please listen to and accept this hymn of mine. //5//

O Sindhu (Indus), flowing, you first meet the Tṛiṣṭāmā (and then) the Susartu, the Rasā and the Śvetā (Swat) and thereafter the Kubhā (Kabul), the Gomatī (Gomal), the Krumu (Kurram) with the Mehatnu; and (finally) you move on in the same chariot with them (i.e. carry their waters with you). //6//

From ***RV*** **10.75.5** it is absolutely clear that the **Ṛigvedic Sarasvatī lay** ***between*** **the Yamunā and Sutlej.** Do we have the Yamunā and the Sutlej in Afghanistan? If not, how can we drag the Ṛigvedic Sarasvatī there?

The *Ṛigveda* (**3.23.4**) also refers to the **Dṛiṣadvatī and Āpayā as the tributaries of the Sarasvatī.** These two rivers are also not there in Afghanistan. These are in Haryana which is a part of India. So, once again, how can the Sarasvatī be placed in Afghanistan?

Further, according to ***RV*** **7.95.2**, as quoted earlier, the **Sarasvatī flowed from the mountains to the sea.** Now, whereas there do exist mountains in Afghanistan, there is **no sea there.** Hence even on this count the case for locating the Ṛigvedic Sarasvatī in Afghanistan is unsustainable.

In view of the foregoing overwhelming pieces of evidence, it is very much to be hoped that the proponents of the view that the Ṛigvedic Sarasvatī lay in Afghanistan will abandon their untenable, outdated thesis.

But then exactly which river in India is to be identified with the Ṛigvedic Sarasvatī? There does exist a river called the Sarasvatī which, starting from the Siwalik foothills, flows through Haryana, and fulfils the most vital of the conditions, viz. its location between the Yamunā and the Sutlej (***RV*** **10.75.5**, quoted earlier; see Fig. 5.2). Passing by Pipli, Kurukshetra and Pehowa, it joins

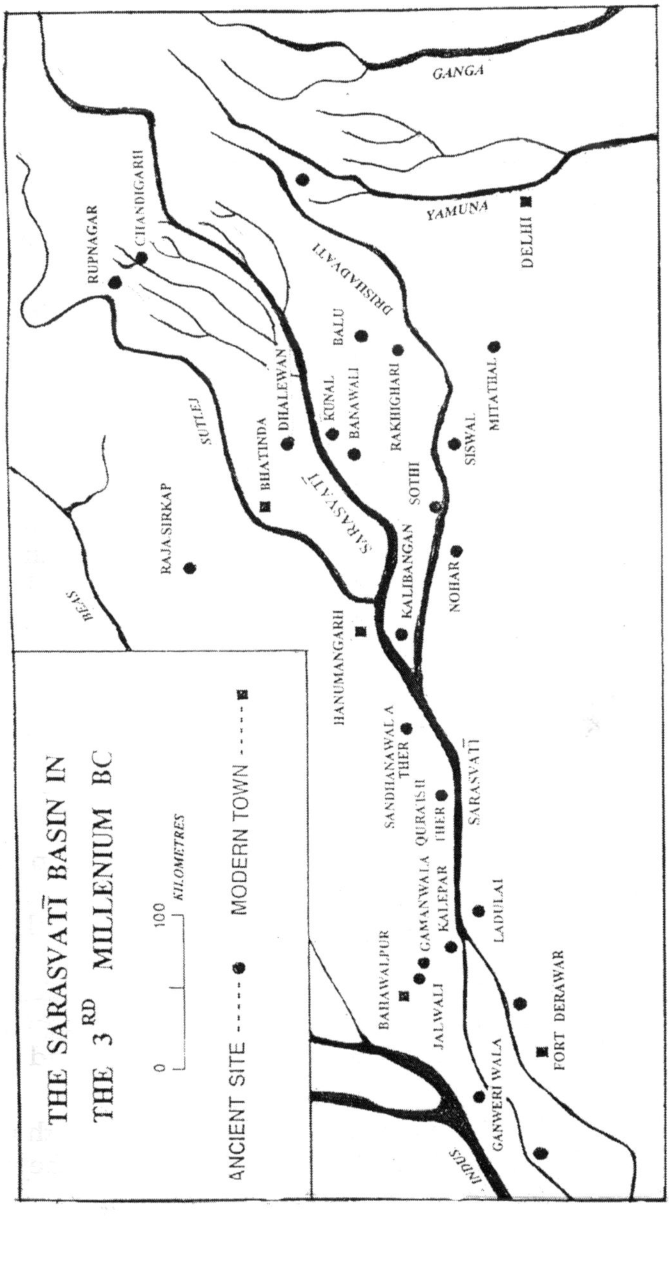

Fig. 5.2 The Sarasvatī Basin in the 3rd millennium BC.

the Ghaggar. Though today it is not a mighty river and dries up beyond Sirsa, its dry bed, sometimes having a width of 5-6 km, can easily be identified all through its ancient lower course – in northern Rajasthan, from where, beyond Suratgarh, it enters Cholistan in Pakistan and moves down through Sindh, to debouch into the Rann of Kachchha (Fig. 5.3). This course is amply clear on the landsat images and thus fulfils the next condition, viz. 'flowing from the mountains to the sea' (***RV*** **7.95.2**, above). And even the third condition, viz. that of having the Dṛiṣadvatī as its tributary (cf. ***RV*** **3.23.4**), is fulfilled. Though no longer an active river, the ancient bed of the Dṛiṣadvatī (now known as the Chautang), can be easily identified at Bhadra, Nohar, etc. It joined the Sarasvatī-Ghaggar combine near Suratgarh (cf. Fig. 5.2).

The foregoing evidence should leave no doubt in the mind of anyone that it is the Sarasvatī-Ghaggar combine in India and **not** the Helmand of Afghanistan, that represents the Ṛigvedic Sarasvatī.

We may now return to the evidence from the Harappan site of Kalibangan which stood on the left bank of the Sarasvatī (now going, of course, by the name of its consort, namely the Ghaggar). Radiocarbon dates indicate that the Mature Harappan settlement over here had to be abandoned around 2000-1900 BC (Lal 1997: 245-46). And, as the hydrological evidence indicates, this abandonment took place on account of the drying up of the Sarasvatī. This latter part is duly established by the work of Raikes, an Italian hydrologist, and of his Indian collaborators. Raikes (1968) has very significantly titled his paper, 'Kalibangan: Death from Natural Causes'. In this context it may be recalled that the **Sarasvatī was a mighty flowing river during the Ṛigvedic times.** Thus, a combined and in-depth study of the archaeological-cum-hydrological-cum-radiocarbon-cum-literary evidence duly establishes that **the *Ṛigveda* must <u>antedate</u> ca. 2000 BC**. By how many centuries, it can be anybody's guess.

We may now also recall the geographical evidence yielded by the famous *Nadi-stuti* hymn (***RV*** **10.75.5 and 6**) which enumerates the rivers familiar to the Ṛigvedic Aryans, surprisingly in **a serial order** from the east to the west. It begins with the Gaṅgā, and mentioning Yamunā, Sarasvatī, etc., ends up with the Indus, adding thereto even the names of its western tributaries such as the Kabul, Gomal and Kurram. It needs to be noted that these latter rivers flow for a major part in eastern Afghanistan. One might now pose a very critical

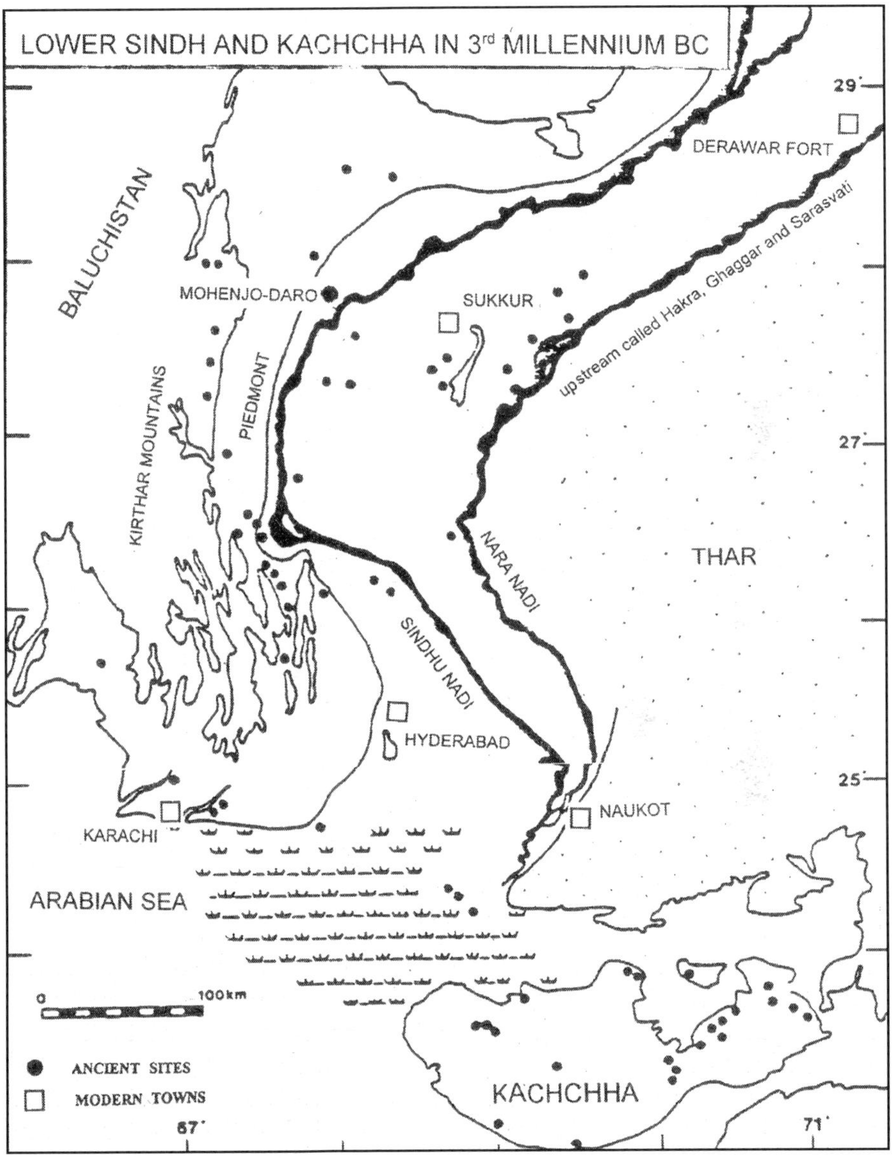

Fig. 5.3 Lower Sindh and Kachccha in the 3rd millennium BC.

question: '**Which archaeological culture/civilization was there in this very area during the pendency of the *Ṛigveda*, namely during the period preceding 2000 BC ? The inescapable answer will have to be: 'The Harappan Civilization!'** (Fig. 5.4).

In other words, the entire foregoing evidence points to a correlation between the *Ṛigveda* and the Harappan Civilization. At the same time, we must take into account the various objections that have been raised from time to time against the Harappan=Vedic equation. Although these have fully been dealt with in my recent book, *The Sarasvati Flows On* (Lal 2002), the position once again needs to be summarized here.

Four major objections have been raised against a correlation between the Vedic and Harappan Civilizations. These are: (i) whereas the Harappan Civilization was urban, the Vedic civilization was that of nomads; (ii) while the horse played an important part in the life of the Vedic people, the Harappans did not use the horse; (iii) likewise, whereas the Vedic people used the spoked wheel in their vehicles, the Harappans were ignorant of the spoked wheel; and finally (iv) whereas according to Max Muller the Vedas were datable to only 1200 BC and the Harappan Civilization was ascribable to the third millennium BC, there was such a wide chronological gap between the two.

In the first place, it is highly misleading to say that the Vedic Aryans were nomads. The *Ṛigveda* itself throws valuable light on the polity of the times, as indicated by the occurrence in it of such terms as ***sabhā* and *samiti* on the one hand and of *samrāṭ, rājan, rājaka,*** etc. on the other. The first two terms clearly refer to assemblies that took vital decisions on matters of public interest. That there did exist these institutions in the Vedic society is clearly borne out by the following verse of the *Ṛigveda* (**9.92.6**):

> *pari sadmeva paśumānti hotā* ***rajā na satyaḥ samitīriyānaḥ*** */*
>
> *somaḥ punānaḥ kalaśāṃ ayāsīt sīdan mṛigo na mahiṣo vaneṣu //*
>
> As the priest seeks the station rich in cattle, **like a true King who goes to great assemblies,**
>
> Soma hath sought the beakers while they cleansed him, and like a wild bull, in the wood hath settled.

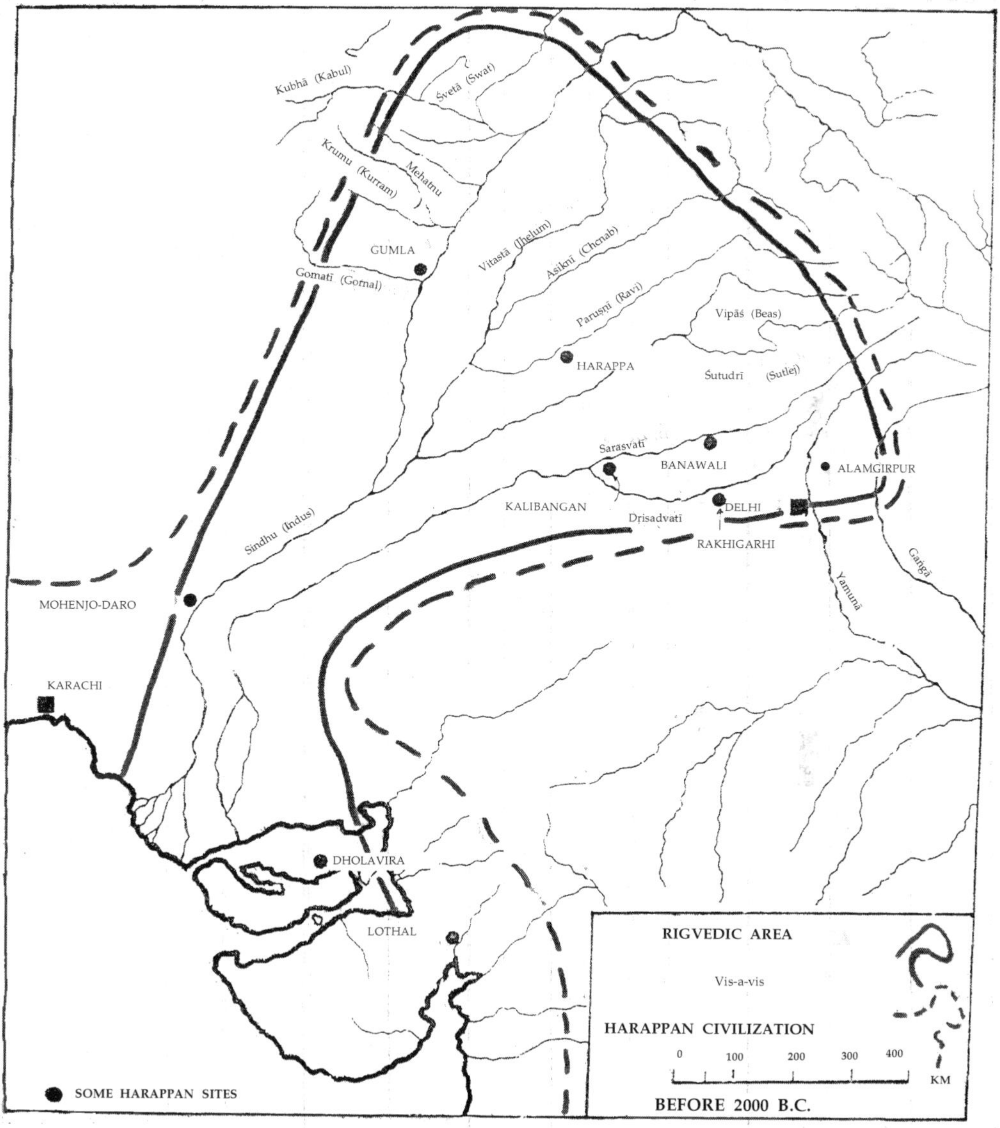

Fig. 5.4 Map showing a correlation between the Ṛigvedic area and the spread of the Harappan Civilization, before 2000 BC.

In the above-noted verse, certain similes are given: viz. that the Soma enters the pitcher (*kalaśa*; not beaker as Griffith translates this word) just as **the king enters the assembly** or a wild buffalo enters the forests. (Sorry, again. Griffith wrongly translates the word *'mahiśa'* as 'bull' and not 'buffalo'.)

The other three terms, viz. ***samrāṭ, rājan* and *rājaka,*** point to a **hierachy** of rulers. In ***RV* 6.27.8, Abhyāvartī Chāyamāna is referred to as a Samrāṭ, whereas in *RV* 8.21.18 Chitra is said to be a mere Rājan and the epithet of other still inferior rulers is Rājaka.**

***RV* 6.27.8:**

dvyāṃ Agne rathino viṁśatim gā vadhūmato maghavā mahyam ***samrāṭ*** /

Abhyāvartī Chāyamāno *dadāti dūṇāśeyam dakṣiṇā pārthavānām* //

Two wagon-teams, with damsels, twenty oxen, O Agni, **Abhyāvartin Chāyamāna,**

The liberal **Sovran**, giveth me. This guerdon of Pṛithu's seed is hard to win from others.

***RV* 8.21.18:**

Chitra id rājā rājakā idanyake *yake Sarasvatīmanu* /

parjanya iva tatanaddhi vṛiṣṭyā sahasramayutā dadat //

Chitra is King, and only kinglings are the rest who dwell beside Sarasvatī.

He, like Parjanya with his rain, hath spread himself with thousand, yea, with myriad gifts.

Does one expect such fine distinction of governance in a nomadic society? That these distinctions were not imaginary but very much real is borne out by the *Śatapatha Brāhmaṇa* (V.1.1.13) when it clearly states:

> *Rājā vai Rājasūyeneṣṭvā bhavati, Samrāḍ Vājapeyena /* ***avaraṁ hi rājyaṁ paraṁ sāmrājyaṁ*** */ Kāmayeta vai Rājā Samārḍ bhavitum avaraṁ hi rājyam paraṁ sāmrājyaṁ / Na Samrāṭ kāmayeta Rājā bhavitum avaraṁ hi rājyaṁ paraṁ sāmrājyaṁ /*

By offering the *Rājasūya* he becomes *Rājan* and by the *Vājapeya* he becomes Samrāj, and **the office of *Rājan* is lower and that of *Samrāj* the higher.** A *Rājan* might indeed wish to become a *Samrāj*, for the office of *Rājan* is lower and of *Samrāj* the higher; but the *Samrāj* would not wish to become a *Rājan* for the office of the *Rājan* is the lower, and that of *Samrāj* the higher.

The mention of armours and forts clearly points to the high-level military strategy of the Aryans. Thus, for example, ***RV* 10.101.8** mentions:

stitch ye [oh gods] **the coats of armour, wide and many, make strong forts as of metal, safe from all assailants.**

In Verse ***RV* 7.15.14**, the devotee further prays that the **fort should have a hundred arms**. Is this the prayer of a nomad?

The Vedic Aryans were well up on the economic front too. As indicated by ***RV* 9.33.6**, they were engaged even in sea-trade.

O Soma, from every side pour forth four seas filled with a thousand-fold riches.

In order to carry out efficiently the sea-trade, the Vedic Aryans used sizeable boats which had as many as **a hundred** (i.e. a very large number of) **oars**. This is indicated by ***RV* 1.116.5:**

anārambhaṇe tadavīrayethāmanāsthāne agrabhaṇe samudre /

yadaśvinā ūhathurbhujyumastam ***śatāritrām nāvamātasthivānsam*** *//*

Ye wrought that hero exploit in the ocean which giveth no support, or hold or station,

What time ye carried Bhujyu to his dwelling, **borne in a ship with hundred oars**, O Aśvins.

Continuing with the topic of transport during Ṛigvedic times, one notes that besides carts there were chariots made from the best wood available for the purpose. It may also be noted that sometimes the chariot was drawn by as many as **four horses**, and the horses were bedecked with ornaments of pearls. Is this the kind of thing one expects in a society alleged to be in a nomadic state ? The relevant verse **(*RV* 1.126.4)** runs as follows:

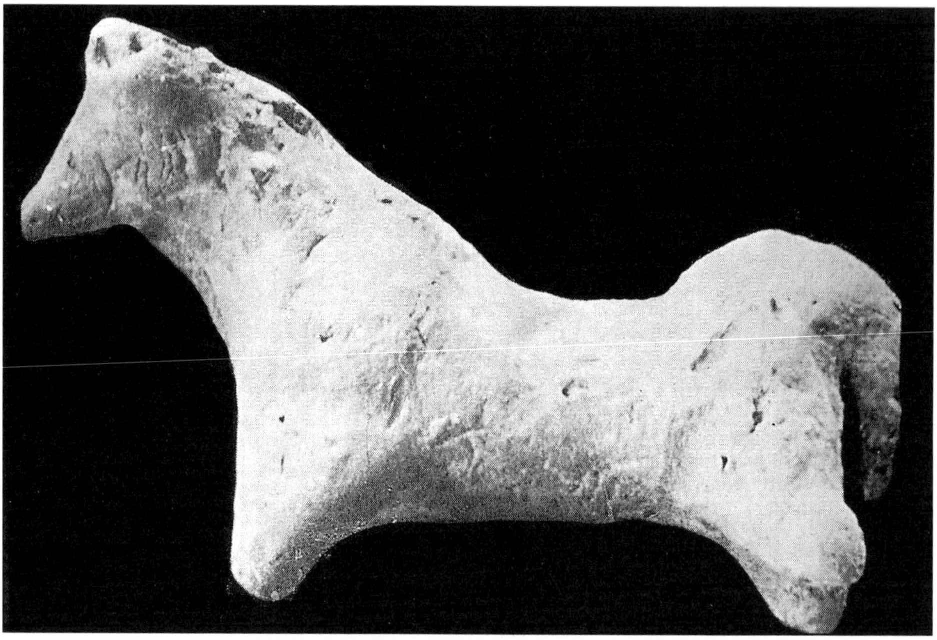

Fig. 5.5 Lothal: Terracotta horse. Mature Harappan.

chatvārinśad daśarathasya śoṇāḥ *sahasrasyāgre śreṇim nayanti /*

madachyutaḥ ***kṛiśanāvato*** *atyān Kakṣīvanta udamṛikṣanta pajrāḥ //*

Forty bay horses of the ten cars' master before a thousand lead the long procession.

Reeling in joy Kakṣīvān's sons and Pajrā's have grounded the **coursers decked with pearly trappings.**

The foregoing few examples should leave no doubt in the mind of anyone that, far from being nomads, the Vedic Aryans had reached a very advanced social, political and economic stage.

Even the much-touted argument about the absence of the horse from the Harappan Civilization has no validity in the light of the new evidence regarding its presence. The noted international authority on the palaeontology of the horse, Sandor Bokonyi of the Archaeological Institute, Budapest, after duly examining

Fig. 5.6 Rakhigarhi: Terracotta wheel. The painted lines represent the spokes. Mature Harappan.

the faunal remains concerned, had declared as far back as 1993, that '**the domestic nature of the Surkotada horse [a Harappan site in Kuchchha] is undoubtful'.** The terracotta model of the horse, found by Mackay in his excavations at Mohenjo-daro, is displayed in the National Museum, New Delhi, and all are welcome to see it. Lothal, the famous Harappan site in Gujarat, has also yielded a **terracotta figures of the horse (Fig. 5.5)**. No less compelling is the faunal evidence from other Harappan sites, like Kalibangan, Rupnagar, etc. (see Sharma, A.K. 1993; Lal 1998: 109-12).

Fig. 5.7 Banawali: Terracotta wheels, showing spokes in low relief. Mature Harappan.

Fig. 5.8 Nausharo: Terracotta female figure, painted. The yellow colour on the ornaments suggests that these were made of gold; the hair is black, while the red on the partition line of the hair indicates the use of vermilion. Ca. 2800-2600 BC.

There is yet another argument that is often advanced against an Aryan-Harappan equation. This relates to the spoked wheel which is referred to in the Vedic texts (e.g. *RV* 1.32.15; 5.13.6; 10.78.4) but is stated to be absent during the Harappan times. **Sites like Banawali, Kalibangan and Rakhigarhi have yielded ample evidence of spoked wheels from the Mature Harappan levels (Figs. 5.6 and 5.7).**

A few scholars, including Sir Mortimer Wheeler, had suggested that the Aryans invaded India around 1500 BC and destroyed the Harappan Civilization, which became **extinct**. This theory has been shown to be completely wrong by a host of scholars, including Professors George F. Dales (1964), Jonathan Mark Kenoyer (1998), Jim G. Shaffer and Diane Lichtenstein (1999) and Lord Colin Renfrew (1988). The last-named even adds (pp. 188 & 190): 'When Wheeler speaks of 'the Aryan Invasion of the Land of Seven Rivers, the Panjab', he has no warranty at all, so far as I can see. If one checks the dozen references in the *Ṛigveda* to the Seven Rivers, **there is nothing in any of them that to me implies invasion..... Despite Wheeler's comments, it is difficult to see what is particularly non-Aryan about the Indus Valley Civilization.'**

All told, therefore, there is a very strong case for the Harappan Civilization having been Vedic, though this equation must await a valid decipherment of the Harappan script.

Fig. 5.9 Harappa: Terracotta figure, greeting with *namaste*. Mature Harappan.

Further, the statement that the Harappan Civilization became 'extinct' has been demonstrated to be completely wrong by the present writer in his book, *The Sarasvatī Flows On: The Continuity of Indian Civilization* (Lal 2002). Indeed, there is hardly any walk of day-to-day life, such agriculture, cooking habits, ornaments, objects of toiletry, personal make-up, games played by adults or children, folk tales, religious practices, etc. in which the carry-overs from the Harappa Culture are not reflected. **Would you believe that the practice of applying vermilion (*sindūra*) to the *māṅga* (medial partition line of the hair on the head) by married Hindu women goes back even prior to the Harappan times (Fig. 5.8), or that the Hindu way of greeting one another has also its root in the Harappan Culture itself (Fig. 5.9)?**

From what has been stated in the preceding pages, it becomes abundantly clear that the Vedas reflect the literary counterpart of the Harappan archaeological complex. And since the Harappan Culture itself had its roots going deep at least into the fifth millennium BC, it would imply that the Sanskrit-speaking Vedic people were already there in this area as early as that. Further, had the Sanskrit-speaking people not been the original inhabitants of this region, we would have got evidence thereof in terms of a substratum language, which we really do not have. The presence of a few Dravidian words in the Vedas can be explained by an ad-stratum and not necessarily by a substratum. As explained elsewhere by the present author (in press), the Harappans came in lateral contact with the Southern Neolithic people who, in all probability, were speakers the Dravidian language.

We now turn to yet another important piece of evidence. The Boghaz Kuei inscription, dating back to the fourteenth century BC, refers to Indra, Mitra, Nāsatya and Varuṇa as witnesses to a treaty between the Mitanni king Matiwaza and the Hittite king Suppiluliuma. There is also the evidence furnished by a text on the training of horses, which uses typical Sanskrit terms like *ekavartana, trivartana,* etc. Further, there are many Indian names in the region going back to ca. seventeenth century BC. After a thorough examination of the entire evidence, the renowned scholar T. Burrow (1955) came to the conclusion: 'The Aryans appear in Mitanni from 1500 BC as the ruling dynasty, which means that **they must have entered the country as conquerors.'** If so, from where could have these conquerors come? **Around 1500 BC there was no other country in the entire world except India where these above-mentioned gods were worshipped. (As we have shown earlier (p. 74), the *Ṛigveda* is datable to at least 2000 BC, if not earlier.) Putting the pieces of the puzzle together, it becomes clear that the immigrants to Anatolia were from no area other than India.** This movement is likely to have taken place along the belt lying south of the Black and Caspian Seas.

The shifting of the 'original homeland' from Sogdiana **to the region now comprising eastern Afghanistan, Pakistan and north-west India should not upset anyone, since the archaeological-cum-literary evidence from this area is much more positive than that from Sogdiana.**

P.S. Having utterly failed to establish the 'Aryan Invasion' theory, the proponents are now taking recourse to that of 'Migration'. While I have dealt with this issue in detail in another paper (Lal, in press), here the following may suffice. Romila Thapar (1989-91) states: 'If invasion is discarded then the mechanism of migration and occasional contacts come into sharper focus. The migrations appear to have been of pastoral cattle-herders who are prominent in the *Avesta* and *Ṛigveda*'. Pressing the point, R.S. Sharma, observes (1999: 77) '...the pastoralists who moved to the Indian borderland came from Bactria-Margiana Archaeological Complex or the BMAC which saw the genesis of the culture of the *Ṛigveda*'. **But can Sharma cite even a single site east of the Indus (which was the main scene of the activity of the Ṛigvedic people) where the remains of the BMAC have been found? In so far as the present author is aware, there is none. Then why indulge in such speculations?**

CHAPTER VI

SHOULD ONE GIVE UP ALL ETHICS FOR PROMOTING ONE'S THEORY?

Please insert the **unexpressed-but-implied verb** in the second part of the following two sentences, in order to make them more explicit:

(i) After the dinner-party was over, George returned home and Raymond to his hotel.

(ii) In the recent cricket final between England and India, the former scored 324 runs whereas the latter 325, thereby winning the match.

You would naturally retort: 'Why must I do that? What is its relevance in a scholarly paper, anyway?' My answer would be: 'Yes, there is. Indeed, very much.'

Since, as a reader of this paper, you are far away, and cannot participate in this (seemingly funny) exercise, I asked some of my colleagues to participate and I give below their response.

In the first case, the completed sentence was: 'After the dinner-party was over, George returned home and Raymond **returned** to his hotel.'

In the second case, the completed sentence was: 'In the recent cricket match between England and India, the former scored 324 runs whereas the latter **scored** 325, thereby winning the match.'

It would thus be seen that the ***unexpressed- but- implied verb*** **in the second part was** ***exactly the same*** **as it was in the first.**

We may now proceed to the real issue. In his paper, 'Ṛgvedic history: poets, chieftains and polities', Michael Witzel, a Sanskritist at Harvard University, writes (1995: 320-21):

> Taking a look at the data relating to the immigration of Indo-Aryans into South Asia, one is struck by the number of vague reminiscences of foreign localities and tribes in the Ṛgveda, in spite repeated assertions to the contrary in the secondary literature. Then, there is the following **direct statement** contained in (the admittedly much later) BŚS [*Baudhāyana Śrautasūtra*], 18.44: 397.9 sqq which has once again been overlooked, **not having been translated yet:** "Ayu went eastwards. His (people) are the Kurū-Pañcāla and Kāśī-Videha. This is the Āyava (migration). (His other people) **stayed at home in the west.** His people are the Gāndhārī, Parśu and Araṭṭa. This is the Amāvasava (group). (Emphasis mine.)

The craze for establishing that there was an Indo-Aryan invasion is so great that any evidence, howsoever fragile, is immediately grasped by all those who are bent on establishing this theory. Thus, R.S. Sharma, a faithful follower of Witzel, comes out triumphantly with the following statement in his book, *Advent of the Aryans in India* (1999: 87-89):

> The latest study of the Ṛgveda history by Michael Witzel, a Sanskritist, gives convincing proof of migration.... More importantly, Witzel produces a passage from the *Baudhāyana Śrautasūtra* which contains 'the most explicit statement of an immigration into the subcontinent.' [It is the same passage as referred to above by Witzel.] The passage contains [Sharma goes on to say] a dialogue between Aila Purūrava and Urvaśī, which refers to the horse, chariot parts, 100 houses and 100 jars of ghee. Towards the end it speaks of the birth of their sons Ayu and Amāvasu, who were asked by their mother to go out. 'Ayu went eastward. His people are the Kurū-Pañcālas and Kāśī-Videhas. This is the Āyava kin group. Amāvasu **stayed in the west.** His people are the Gandharas, the Pārśavas and the Araṭṭas. This is the Amāvasava kin group.'

It may be noted that Sharma parrots what Witzel had said, namely 'Amāvasu **stayed in the west**'. One does not know if Sharma ever cared to examine whether the translation of Witzel was correct or not. Or, why should he bother about it, since here was the most welcome chance for him to cite a Sanskritist from the West to support the thesis of Aryan Invasion of which he himself is one of the most vocal supporters in India?

[१८ । ४४, ४५ ।] बौधायनश्रौतसूत्रम् । ३९७

अपररात्रागच्छेयुस्तदाग्नना स्यां । न त्वा नग्नं पश्येयमिति । सर्वमेवैतद्भगवति सुकरमिति होवाच । कथा त्वपि[१] जाया पतिं नग्नं न पश्यतीत्यन्तर्वासं वसीथा[२] इति होवाचानग्नो भवेति[३] । तथा सहोवाचान्तर्वासं वसानः । सा ह स्म जाताञ्जातानेव पुत्रानपविध्यति । तान्ह ह राजोवाच पुत्रकामा ह वै भगवति वयं मनुष्याः स्मो जाताञ्जातांस्तु त्वमपविध्यसीति । सा होवाच पर्यवेतराजयो भवन्ति क्षीणायुषो ऽन्ये । भूयः प्रियं करवावहा इति । सायुं[४] चामावसुं च जनयां चकार । सा होवाचेमौ बिभृतेमौ[५] सर्वमायुरेष्यत इति । प्राङायुः प्रवव्राज । तस्यैते कुरुपञ्चालाः[६] काशिविदेहा इत्येतदायवं प्रव्राजं । प्रत्यङ्ङमावसुस्तस्यैते गान्धारय[७] स्पर्शवो ऽरात्ता इत्येतदामावसवम्[८] । ॥ ४४ ॥

अथो हास्या एषा पूर्वचित्तिरप्सरा स्वसा बभूव । सा हेर्ष्यां चक्रे ज्योग् वै मे स्वसा मनुष्येष्ववात्सीद्धन्तैनामच्छाया-

---

१ त्वपि L, त्वपि the other MSS.

२ वसीथा L Be (sec. m.) वासीत्याथ the other MSS.

३ भवति MSS. exc. L, which has वेन च वामनो भवतीति ।

४ प्रायुं Be (sec. m.).

५ बिभृतौ॰ Be L ; the other MSS. om. these words.

६ ॰पांचालाः L B.

७ The proper names given acc. to Be (sec. m.) ; the other MSS. गांधारा (or री) पयस्पर्शवोद्दारा (or त्य) सा (or वा) ।

८ इत्येतन्मावसुम् or इत्येतन्मामवसुम् or इत्येतन्मामावसम् MSS.

Fig. 6.1 Photocopy of the relevant Sanskrit text from the *Baudhāyana Śrautasūtra*.

To come to the Sanskrit text itself, it may be noted that Witzel states that it was never translated before and takes pride on the fact that he was translating it for the first time. But have a look at the mess he has made of it. Witzel quotes in the footnote (no. 27, on p. 321) the original text, which runs as follows:

Prāṅāyuḥ pravavrāja. tasyaite Kurū-Pañcālāḥ Kāśī-Videhā ity. etad Āyavam. pratyaṅ Amāvasus tasyaite Gāndhārayas Parśavo "araṭṭā ity. etad Āmāvasavam.

Referring back to the two model sentences we took up at the beginning of this paper, it would be absolutely clear that the ***unexpressed-but-implied verb*** **in the second part of the above-mentioned sentence has got to be the** ***same*** **as in the first part. In other words, the second part must be rendered as:** ***pratyaṅ Amāvasuḥ*** ***pravavrāja*** ***tasyaite Gāndhārayas Parśavo "rāṭṭā ity, etad Āmāvasavam.***

The correct English translation of the entire given text would, therefore, be: **"Āyu migrated eastwards. [The word '*pravavrāja*' would mean 'migrated' and not just 'went'.] His (people) are the Kuru-Pāñchālas and Kāśī -Videhas. This is the Āyava (migration). Amāvasu *migrated* westwards. His (people) are the Gāndhārī, Parśu and Āraṭṭa. This is the Amāvasava (migration)."** To avoid any confusion, a photocopy of the original text, as published by Caland (1907, p. 397), is given here (Fig. 6.1).

What, then, the text really says is that (from some *intermediary* region) Āyu migrated eastwards and Amāvasu *migrated* westwards. There was no question of the latter *staying at home in the west*. In other words, the area of *parting* must have been somewhere between the Gāndhāra region on the west and the Kuru region on the east. Since the Gāndhāra region lay on either side of the Kabul river, immediately above its junction with the Indus, and the Kuru region (modern Kurukshetra) is in Haryana in India, the area from where these eastward and westward migrations took place must have been, in all likelihood, somewhere in the pre-Partition Punjab.

At the end, it may be repeated that the **interpolation** of the words **'stayed at home'** by Witzel in his translation of the second part of the text is **absolutely *wrong and wilful.*** Since Witzel was all out to prove an 'immigration of Indo-Aryans **into** South Asia' (cf. his opening sentence, above), he **deliberately and consciously used the words "*stayed at home*" – which is quite *unethical.* May it be hoped that such practices are not resorted to any more?**

APPENDICES

1. Wood-charcoal from Some Harappan Sites in Haryana and Panjab
2. The *Aśva* (Horse)
3. The *Vṛiṣabha* (*Bos indicus*)

APPENDIX 1

WOOD-CHARCOAL FROM SOME HARAPPAN SITES IN HARYANA AND PANJAB*

The ancient settlements in the valley of the Sarasvatī must have been founded on environmental resources where the ecological conditions allowed an effective exploitation of the land for large-scale agriculture. Studies of wood-charcoal from Kunal, Banawali and Rohira indicate the presence in the region of a tropical, moist-deciduous forest in the third millennium BC. We have evidence of increased discharge down the plains of Panjab and Haryana. Flooding in the Sarasvatī, Sutlej and other rivers would have inundated considerable areas as a result of which thick ground vegetation and deciduous forests along the water-courses would have retained high moisture-content in the past.

Factual evidence of charcoal remains indicate the presence of the following species in the region: *gūlara* (*Ficus glomerata*); *bela* (*Aegle marmelos*); *tūnā* (*cedrela toona*); *palāśa* (*Butea monosperma*); *banklī* (*Anogeissus latifolia*); *śīśama* (*Dalbergia sissoo*); *dhera* (*Diospyros* sp.); *sehajana* (*Moringa pterigosperma*); *dahiā* (*Streblus asper*); *jāmuna* (*Syzygium* sp.); *jhāu* (*Tamarix diocia*); *khaira* (*Acacia catechu*) and many other moisture-loving deciduous trees/plants, along with the groves of bamboo (*Bambusa* sp.) by the side of water-streams. The quality timbers of pine (*Pinus roxburghii*), deodar (*Cedrus deodara*), teak (*Tectona grandis*) were transported from distant forests for their specific use. *Neema* (*Azadirachata indica*) was grown as an avenue tree. Charcoals of *nībū* (*Citrus* cf. *limon*), *śahatūta* (*Morus*

---

* Contributed by Dr. K.S. Saraswat

alba), *anāra* (*Punica granatum*), *harasingāra* (*Nyctanthes arbortristis*) and grape-vine (*Vitis vinifera*) reflect on the horticultural activities ofc the early settlers. Entry of the xeric elements of the arid zones, such as *jhandī/śamī/chhonkara* (*Prosopis cineraria*), *karīla* (*Capparis decidua*), *pīlū* (*Salvadora oleoides*), *farash* (*Tamarix arteculata*), *phok* (*Ephedra* sp.), possibly *babūla* (*Acacia nilotica*), etc. would have taken place with a variety of human activities, such as the clearance of the land for agricultural operation. Since the land has been subjected to reckless exploitation over the millennia, the result is the present arid conditions in the region during the sub-recent times.

The Table A1.1 that follows gives details of the charcoal samples that have been identified.

A1.1 WOOD CHARCOAL FROM SOME HARAPPAN SITES IN HARYANA AND PANJAB

Sl. No.	Taxa	Kunal (Haryana) Early Harappan		Banawali (Haryana)		Rohira (Panjab)	
		IB ca. 2850-2600 BC	IC ca. 2600-2500 BC	Early Harappan ca. 2750-2500BC	Mature Harappan ca. 2500-2000BC	Early Harappan	Mature Harappan
1.	*Acacia nilotica/catechu* (Babul/Khair)	■	■	■	■	■	■
2.	*Aegle marmeios* (Bel)		■				■
3.	*Albizia cf. lebbeck* (Siris)			■	■		
4.	*Anogeissus latifolia* (Dhaura/Bankli)				■		
5.	*Azadirachta indica* (Neem)		■		■		
6.	*Bambusa sp.* (Bamboo)	■	■		■		
7.	*Bombax ceiba* (Semal/Senwal)						■
8.	*Boswellia serrata* (Salai)	■	■		■		

Sl. No.	Taxa	Kunal (Haryana) Early Harappan		Banawali (Haryana)		Rohira (Panjab)	
		IB ca. 2850-2600 BC	IC ca. 2600-2500 BC	Early Harappan ca. 2750-2500BC	Mature Harappan ca. 2500-2000BC	Early Harappan	Mature Harappan
9.	*Butea monosperma* (Palash)	■	■				■
10.	*Caeselpinia sp.* (Karangi/Kanderi)	■					
11.	*Calotropis sp.* (Aak/Madar)		■				
12.	*Cannabis sativa* (Bhang)	■	■				
13.	*Capparis decidua* (Kareel)			■	■	■	
14.	*Capparis sepiaria/ horrida* (Heens/Jhiri)		■				
15.	*Cedrela toona* (Tun)					■	■
16.	*Cedrus deodara* (Deodar)		■		■	■	
17.	*Citrus cf. limon* (Neebu/Nimbu)		■		■		

Sl. No.	Taxa	Kunal (Haryana) Early Harappan		Banawali (Haryana)		Rohira (Panjab)	
		IB ca. 2850-2600 BC	IC ca. 2600-2500 BC	Early Harappan ca. 2750-2500BC	Mature Harappan ca. 2500-2000BC	Early Harappan	Mature Harappan
18.	*Dalbergia cf. sissoo* (Shisham)		■		■		■
19.	*Datura sp.* (Dhatura)		■				
20.	*Diospyros sp.* (Dhera)			■	■		
21.	*Ephedra sp.* (Phok)		■				
22.	*Feronia limonia* (Kaitha)	■					
23.	*Ficus glomerata* (Gular)	■	■	■	■	■	
24.	*Ficus religiosa* (Pipal)		■		■		■
25.	*Lawsonia inermis* (Heena/Mehandi)	■		■		■	
26.	*Manilkara hexandra* (Khirni)					■	

Sl. No.	Taxa	Kunal (Haryana) Early Harappan		Banawali (Haryana)		Rohira (Panjab)	
		IB ca. 2850-2600 BC	IC ca. 2600-2500 BC	Early Harappan ca. 2750-2500BC	Mature Harappan ca. 2500-2000BC	Early Harappan	Mature Harappan
27.	*Moringa pterigosperma* (Sehjan)	■	■				
28.	*Morus alba* (Shahtoot/Mulberry)		■	■	■		
29.	*Nyctanthes arbortristis* (Harsingar)						■
30.	*Phoenix sp.* (Date)			■			
31.	*Pinus roxburghii* (Chir pine)	■					
32.	*Prosopis cineraria* (Jhandia)			■	■		■
33.	*Punica granatum* (Anar/Dadim)				■		■
34.	*Salvadora oleoides* (Pilu)				■		
35.	*Syzgium cf. cumini* (Jamun)	■	■		■		■

Sl. No.	Taxa	Kunal (Haryana) Early Harappan		Banawali (Haryana)		Rohira (Panjab)	
		IB ca. 2850-2600 BC	IC ca. 2600-2500 BC	Early Harappan ca. 2750-2500BC	Mature Harappan ca. 2500-2000BC	Early Harappan	Mature Harappan
36.	*Tamarix arteculata* (Farash)	■			■		
37.	*Tamarix dioica* (Jhau)	■				■	■
38.	*Tectona grandis* (Teak)				■	■	
39.	*Vitis vinifera* (Grape)		■			■	■
40.	*Ziziphus sp.* (Ber/Jujube)		■	■	■		■

APPENDIX 2

THE *AŚVA* (HORSE)

The horse had a very important role to play in the life of the Ṛigvedic people. It has been referred to in many hymns of the *Ṛigveda*, some of which are quoted below. In ***RV*** **4.32.17**, the devotee prays to Indra for the grant of a thousand well-trained steeds:

> ***Sahasram vyatīnām*** *yuktānāmindramīmahe /*
>
> *Śatam somasya khāryaḥ //*
>
> To Indra for a **thousand steeds** well-trained and fleet of foot we pray,
>
> And hundred jars of Soma juice.

In praise of Agni, it has been stated, in ***RV*** **5.4.11**, that–

> *Yasmai tvam sukṛite jātaveda u lokamagne kṛiṇavaḥ syonam /*
>
> ***Aśvinam*** *sa putriṇam vīravantam gomantam rayim naśate svasti //*
>
> The pious man, O Jātavedas Agni, to whom thou grantest ample room and pleasure,
>
> Gaineth abundant wealth and sons and **horses,** with heroes and with kine for his well-being.

Quite often the number of horses received as gift could be immense. So says ***RV*** **8.46.22:**

Ṣaṣṭim sahasrāśvyasyāyutāsanamuṣṭrānāṁ viṁśatim śatā /

Daśa śyāvīnāṁ śatā daśa tryaruṣīṇāṁ daśa gavāṁ sahasrā //

Steed sixty thousand and ten thousand kine and twenty hundred camels I obtained;

Ten hundred brown in hue, and other ten red in three spots; in all, ten thousand kine.

Sometimes the poet takes a flight of imagination and compares a star-studded night-sky with a dark-coloured horse adorned with pearls and gold. This has been brought out beautifully in ***RV*** **10.68.11**, as follows:

Abhi śyāvam na kṛiśanebhiraśvam nakṣatrebhiḥ pitaro dyāmapiṁśan /

rātryāṁ tamo adadhurjyotirahanbṛihaspatirbhinadadrim vidagdāḥ //

Like a dark steed adorned with pearl, the Fathers have decorated heaven with constellations.

They set the light in day, in night the darkness. Bṛihaspati cleft the rock and found the cattle.

But the exact identification of the Ṛigvedic horse has been a matter of debate. Was it *Equus caballus* or *Equus hemionus* or *Equus asinus* ? Basing their argument on a phonological equation between the Sanskrit word *aśva* with the Latin word *asinus* and the English word *ass*, some scholars (e.g. Bhagwan Singh 1995: 57) have argued that the Ṛigvedic *aśva* may have been an ass instead of a horse.

But that is not the end of the debate. There is a hymn in the *Ṛigveda* which indicates that **the horse known to the Ṛigvedic people had 17 pairs of ribs instead of 18 pairs as found in most of the modern species (barring a few to which we shall refer shortly). This sets the Vedic horse apart from the other contemporary species.** The relevant Ṛigvedic verse **(1.162.18)** runs as follows:

chatustrinśad vājino devabandhorvaṅkrīraśvasya svadhitiḥ sameti /

achchhidrā gātra vayunā kṛiṇota paruṣparuranughuṣyā vi śasta //

The **four-and-thirty ribs of the swift Charger, kin to the Gods, the slayer's hatchet pierces,**

Cut ye with skill, so that the parts be flawless, and piece by piece declaring them dissect them.

Should one ignore the mention of 34 ribs in the foregoing Vedic text as being accidental or should one even charge the Vedic *Ṛiṣis* concerned with utter ignorance of the anatomy of horses? But the fact that even today there are instances of horses having 17 pairs of ribs must set aside such wishful thinking. We quote below from a communication from Dr. Kevin Seymour, Assistant Curator, Palaeobiology, Royal Ontario Museum.

> The number of ribs in mammals corresponds to the number of thoracic vertebrae. For horses, the usual number is 18 thoracic vertebrae, and therefore 18 pairs of ribs (36 ribs in all). Some living horses have 19 thoracic vertebrae (and hence 38 ribs), and rarely 17 has been recorded too (and hence 34 ribs). We have a quarter horse mounted in our gallery with 18 pairs of ribs, and **a Shetland pony mounted with 17 pairs of ribs. So there is variation even in the living species.**
>
> Although knowing that even in the living horse there is a variation, then **it is also quite possible that <u>at least SOME hipparions had 17 pairs of ribs</u>.**

Besides the Shetland Pony referred to above, there is yet another species, namely **the Spanish Barb, which too has 17 pairs of ribs**. In this context, we quote from Bent's Old Fort NHS, The Learning Page, as follows:

> The Spanish Barb is a rare breed of horse that traces its ancestry to horses brought to North America by Spanish explorers during the 16th century. In this country [i.e. USA], the Spanish Barb was most prevalent in the southeast and southwest, both areas primarily settled by the Spaniards.
>
> These horses formed the basis for the wild mustang herds of the American West. They were the breed of choice among early Spanish and Mexican cowboys. The barbs are of great historical, and "rare breed" value. **The barb has the oldest history of any horse. It is a true and separate breed, with a short back, only 5 lumbar vertebrae rather than 6, <u>and 17 pairs of ribs instead of 18 like other horses</u>.**

The foregoing data must set one thinking about the identity of the Ṛigvedic horse. If indeed it was not *Equus Caballus* Linn. but belonged to another, *indigenous,* species, the argument that the Ṛigvedic horse was of Central Asian origin and hence the Aryans must have come from that region loses all its force.

APPENDIX 3

THE *VṚIṢABHA* (*BOS INDICUS*)

The word *Vṛiṣabha,* meaning the bull, occurs in many hymns of the *Ṛigveda.* The purpose of the present exercise is to find out which particular species of the bull was implied in these references: *Bos indicus* (*Zebu*) or *Bos taurus* ? If it is found to be *Bos indicus,* what can be the possible implications of such an identification?

Here are some of the verses concerned. In ***RV*** **1.94.10** it is the loud roar of the bull that has been emphasized. It is a hymn dedicated to Agnī:

> *Yadayukthā aruṣā rohitā rathe vātajūtā* ***vṛiṣabhasyeva te ravaḥ*** /
>
> *Ādinvasi vanino dhūmaketunāgne sakhye mā riṣāmā vayam tava* //

> When to thy chariot thou hadst yoked two red steeds and two ruddy steeds, wind-sped, **thy roar was like a bull's.**
>
> Thou with smoke-bannered flame attackest forest trees. Let us not in thy friendship, Agni, suffer harm.

In a hymn dedicated to Indra, it is the ferocity/anger of the bull that is highlighted. The relevant verse, ***RV*** **6.46.4,** runs as follows:

> *Bādhase janān* ***vṛiśabheva manyunā ghṛiṣau*** *mīḷaha ṛichīṣama* /
>
> *asmākam bodhyavitā mahādhane tanūṣvapsu sūrye* //

Fig. A3.1 Gold coin of the Kushan king Kanishka (1st century AD), showing on the obverse the king himself and on the reverse Śiva with his *vāhana* (mount), viz. the *vṛiṣabha* (humped bull) behind him.

Ṛichishama, thou **forcest men as with a bull, with anger, in furious fray.**

Be thou our Helper in the mighty battle fought for sunlight, water, and for life.

In ***RV*** **7.101.1**, the loud rumbling of the clouds, before they pour down the rains, has been compared to the bellowing of the bull.

Tisro vāchaḥ pra vada jyotiragrā yā etadduhre madhudoghamūdhaḥ /

sa vatsam kṛiṇvan garbhamoṣadhīnām sadyo jāto ***vṛiṣabho roravīti*** *//*

Speak forth three words, the words which light precedeth, which milk this udder that produceth nectar.

Quickly made manifest, **the bull hath bellowed,** engendering the germ of plants, the Infant.

It is no less important to note that the **bull is associated with Rudra.** This is very clear from many hymns, of which one is quoted below, namely ***RV*** **2.33.15:**

Evā ***babhro vṛiṣabha*** *chekitāna yathā deva na hṛiṇīṣe na haṁsi /*

havanaśrunno ***Rudreha*** *bodhi bṛihadvadema vidathe suvīrāḥ //*

O **twany Bull**, thus showing forth thy nature, as neither to be worth, O God, nor slay us,

Here, **Rudra**, listen to our invocation. Loud may we speak, with heroes, in assembly.

In later mythology Rudra is identified with Śiva. And it is important to note that the vehicle of Śiva is invariably the *vṛiṣabha.* The *Śivamahāpurāṇa* (Sanskrit text by Śrīkṛiṣṇa Dvaipāyana, translated into Hindi by Pandit Jwala Prasad Misra and published by Khemraj Shrikrishnadas, Bombay, 1988), has a section called Chaturthī Kotirudrasaṁhitā. Its Chapter 35 is devoted to the enumeration of a thousand names of Lord Śiva (*Śivasahasranāmavarṇanam*), which a devotee is enjoined to recite every day to propitiate the Lord. Verse No. 6 of this Chapter (p. 846 of the text just referred to) runs as follows:

Viśvarūpo virūpākṣo vāgīśaśśuchisattamaḥ /

Sarvapramāṇasaṁvādī vṛiṣāṅkaḥ ***Vṛiṣavāhanaḥ*** *//*

The term ***Vṛiṣavāhanaḥ*** means **'one who has the bull as his vehicle'.**

It is not a religious book alone that shows the association of the bull with Śiva. The same is the case in secular literature. For example, take the *Raghuvaṁśa,* by the world-renowned Sanskrit poet, Kalidasa. Its Second Canto deals with the story of King Dilip who was enjoined by Sage Vaśiṣṭha to take out his cow for grazing and protect her from all dangers. At one point the king came face-to-face with a lion who wanted to eat up the cow. In order to protect the cow, the king wanted to shoot the lion with his arrows. On seeing this, the lion told the king, in human voice, that, even if discharged, the king's arrow would not be able to harm him (the lion). The lion further added (Nandargikar, G.R. (ed.) 1971, *The Raghuvaṁśa of Kalidasa,* p. 48, Delhi: Motilal Banarsidass):

Kailāsagauram ***vṛiṣamārurukṣoḥ*** *pādārpaṇānugrahapūtapṛiṣṭham /*

Avehi mām ***kiṁkaramaṣṭamūrteḥ*** *Kumbhodaram nāma Nikumbhamitram //*

Fig. A3.2 Stone sculpture from Warangal, Andhra Pradesh, datable to the 12th century, showing dancing Śiva along with his *vāhana* (mount), viz. the *vṛiṣabha* (humped bull).

Be it known to thee, O King, that my name is Kumbhodara, a friend of Nikumbha; and I am the servant of the **eight-formed God (Śiva)**, and my back is sanctified by the favour of placing his footsteps, **when wishing to mount his <u>bull</u> white as the mount Kailāsa.**

That Śiva invariably had/has the bull as his mount is fully borne out by his portrayal on coins and sculptures. Here we produce two examples, one of a coin of the Kushan king Kanishka (AD 78, Fig. A3.1) and another of a sculpture of the 12th century AD (Fig. A3.2), in both of which Lord Śiva is depicted with his *vāhana* (vehicle), viz. the bull.

Now begins the most crucial part of the discussion. As would be seen from the representation of the bull on the coin as well as the sculpture, it is the **humped bull** that is depicted. That would at once mean that **the *vṛiṣabha* in the entire Sanskrit literature, from the *Śivamahāpurāṇa* and *Raghuvaṁśa* back to the *Ṛigveda* was none other than the <u>humped bull.</u>**

We then turn our attention to the significance of the humped bull (*Bos indicus/Zebu*) in the present context. **The faunal remains of this particular animal have been found at Mehrgarh right from the beginning of the settlement, i.e. from as far back as the seventh millennium BC. It continues all through and becomes the hallmark of the Mature Harappan seals which**

Fig. A3.3 Steatite seal from Mohenjo-daro, showing a humped bull (*vṛiṣabha*). Mature Harappan.

are datable to ca. 2600-2000 BC (Fig. A3.3). Further, scholars are of the view that the humped bull (*Bos indicus*) was indigenous to the Indo-Pakistan subcontinent and not an import from elsewhere. We have also seen that the *Ṛigveda* is definitely datable to a period prior to 2000 BC. By how many centuries? It is anybody's guess (see above, p. 74). **In other words, since the *Vṛiṣabha* (humped bull/ *Bos indicus*), indigenous to the Indo-Pakistan subcontinent, is an animal *par excellence* in the *Ṛigveda* as well, the Ṛigvedic people must have also been indigenous to this very area. Such a situation, once again, sets at naught the theory of 'Aryan Invasion/Migration'.**

BIBLIOGRAPHY

Bor, N.L. 1960. *The Grasses of Burma, Ceylon, India and Pakistan.* Oxford.

Brandis, D. 1906. *Indian Trees.* London. (Dehradun: Bishen Singh Mahendra Pal Singh, Reprint, 1971).

Brandis, D. 1874. *The Forest Flora of North-west and Central India.* London. (Dehradun: Bishen Singh Mahendra Pal Singh, Reprint 1972).

Bryant, Edwin. 2001. *The Quest for the Origins of Vedic Culture.* Oxford: Oxford University Press.

Burrow, T. 1955 (Reprint). *The Sanskrit Language.* Faber and Faber.

Caland, W. (ed.). 1907. *The Baudhāyana Śrautasūtra Belonging to the Taittirīya Saṁhitā.* Vol. II, XVIII. 44. Calcutta: Asiatic Society.

——. 1931 (reprinted 1982). *Pañchaviṁśa Brāhmaṇa.* Calcutta: Asiatic Society.

Casal, J.-M. 1961. *Fouilles de Mundigak.* 2 vols. Paris: Libraire C. Klincksieck.

Chowdhury, K.A. and S.S. Ghosh. 1951. Plant Remains from Harappa 1946. *Ancient India,* 7: 3-19.

Collet, H. 1902. *Flora Simlensis.* London (3rd Reprint, Dehradun: Bishen Singh Mahendra Pal Singh).

Cope, Thomas A. 1982. Poacede. In Nasir, E. and S.I. Ali (eds.), *Flora of Pakistan.* Karachi.

Costantini, L. 1979. Wood Remains from Shahr-i-Sokhta: A Source of Information for the Ancient Environment and Technology in Protohistoric Sistan. In M. Taddei (ed.), *South Asian Archaeology 1977,* Vol. 1: pp. 87-127. Naples.

Dales, G.F. 1964. The Mythical Massacre at Mohenjo-daro. *Expedition* 6(3): 36-43.

De Candolle A. 1886. *Origin of Cultivated Plants.* New York (Reprint 1959).

Demoule, Jean-Paul. 1980. Les Indo-Europeans: Ont-ils existe? *L'Histoire* 28: 108-20.

Durante, S. 1975. The Utilization of *Xancus pyrum* L. at Shahr-i-Sokhta: Further Evidence for Cultural Relations Between India and Iran in 4th-3rd Millennium BC. Paper presented at 2nd International Conference on South Asian Archaeology, Paris.

——. 1977. L'utilizzazione delle conchiglie marine importate a Shahr-i-Sokhta. La natura dello scambio economico tra l'Iran interno e le dell'Oceano Indiano. In *La Citta Bruciata nel Deserto Salato*: 214-28. Venezia.

Duthie, J.F. 1903-29. *Flora of Upper Gangetic Plain and of the Adjacent Siwaliks and Sub-himalayan Tracts,* 3 vols., Calcutta.

Friedrich, Paul. 1970. *Proto-Indo-European trees: The arboreal system of a prehistoric people.* Chicago: University of Chicago Press.

Gamble, J.S. 1902. *A Manual of Indian Timbers.* London. (Dehradun: Bishen Singh Mahendra Pal Singh, 2nd Reprint, 1972).

Gamkrelidze, Thomas V. and Vjacelav V. Ivanov. 1995. *The Indo-European and the Indo-Europeans.* Trends in Linguistic Studies and Monographs 80. Berlin: Mouton and Gruyter.

Gimbutas, M. 1966. Proto-Indo-European Culture: The Kurgan Culture During the Fifth, Fourth and Third Millennia BC. In George Cordona et al. (eds.) *Indo-European and Indo-Europeans,* pp. 155-97. Philadelphia: University of Pennsylvania Press.

——. 1997. *The Kurgan Culture and the Indo-Europeanization of Europe.* Ed. Miriam Dexter and Karlene Jones-Bley. Washington, D. C.: Institute for the Study of Man.

Griffith, R.T.H. 1995 (reprint). *The Hymns of the Ṛigveda.* Delhi: Low Price Publications.

Gupta, S.P. 1996. *The Indus-Sarasvatī Civilization: Origins, Problems and Issues.* Delhi: Pratibha Prakashan.

Habib, Irfan and Faiz Habib. 1991-92. The Historical Geography of India, 1800-800 BC. *Proceedings, Indian History Congress: 52nd Session,* pp. 72-97.

Haines, H.H. 1922. *The Botany of Bihar and Orissa,* Part III. London. (Dehradun: Bishen singh Mahendra Pal Singh, Reprint 1978.)

——. 1924. *The Botany of Bihar and Orissa,* Part V. London. (Dehradun: Bishen Singh Mahendra Pal Singh, Reprint 1978.)

Hemphill, B.E., J.R. Lukacs and K.A.R. Kennedy. 1991. Biological Adaptations and Affinities of Bronze Age Harappans. In R. H. Meadow (ed.) *Harappa Excavations 1986-1990,* pp. 137-82. Madison, Wisconsin: Prehistory Press.

Hooker, J.D. 1874. *The Flora of British India,* Vol. II. Ashford, Kent: L. Reeve and Co. (Indian Reprint , Dehradun and Delhi, 1973.)

——. 1878. *Flora of British India*, Vol. V. Ashford, Kent: L. Reeve and Co. (Indian Reprint, Dehradun and Delhi, 1973).

Hooker, J.D. 1880. *The Flora of British India*, Vol. VII. Ashford, Kent: L. Reeve and Co. (Indian Reprint, Dehradun and Delhi, 1973.)

Hyam, R. and R. Pankhurst. 1995. *Plants and their Names*. Oxford: Oxford University Press.

Index Kewensis: An Enumeration of Genera and Species of Flowering Plants, Vol. I. Oxford: The Clarendon Press. 1895.

Jarrige, Catherine, et al. 1995. *Mehrgarh: Field Reports 1974-1985 (From Neolithic Times to Indus Civilization)*. Karachi: Department of Culture and Tourism, Government of Sindh, Pakistan, in collaboration with French Ministry of Foreign Affairs.

Jones, William. 1788. The Third Anniversary Discourse. *Asiatick Researches*, Vol. I, pp. 415-31.

Joshi, J.P. 1990. *Excavation at Surkotada 1971-72 and Explorations in Kutch*. New Delhi: Archaeological Survey of India.

Kenoyer, Jonathan Mark. 1998. *Ancient Cities of the Indus Civilization*. Karachi: Oxford University Press and American Institute of Pakistan Studies.

Kitamura, S. 1960. *Flora of Afghanistan*. Kyoto University, Japan.

Komorov, V.L. 1948. *Flora URSS*, XIII: 385. Editio Academiae Scientiarum URSS, Mosqua, Leningrad.

Krell, Kathrin S. 1998. Gimbutas' Kurgan-PIE Homeland Hypothesis: A Linguistic Critique. In Roger Blench and Mathew Spriggs (eds.) *Archaeology and Language*, II: 267-89. London: Routledge.

Lace, J. H. 1891. A Sketch of the Vegetation of British Baluchistan, with Descriptions of New Species. *Jour. Linn. Soc.*, XXVIII: 288-312.

Lal, B.B. 1970. Some Observations on the Harappan Script. In L. Chandra et al. (eds.). *India's Contribution to World Thought and Culture*, pp. 189-202. Madras.

——. 1974. *Has the Indus Script been Deciphered ? An Assessment of Two Latest Claims*. Shimla: Indian Institute of Advanced Study.

——. 1983. Reading the Indus Script. *Indian and Foreign Review*, 20 (13): 33-36.

——. 1997. *The Earliest Civilization of South Asia*. New Delhi: Aryan Books International.

——. 1998. *India 1947-1997: New light on the Indus Civilization*. New Delhi: Aryan Books International.

——. 2002. *The Sarasvatī Flows On: The Continuity of Indian Culture*. New Delhi: Aryan Books International.

——. (In press). Aryan Invasion of India: Perpetuation of a Myth. In E. Bryant and L. Patton (eds.) *Evidence and Evocation in Indian History*. London: Curzon.

Lamberg-Karlovsky, C. 1988. Indo-Europeans: A Near-Eastern Perspective. *Quarterly Review of Archaeology* 9 (1): 1-10.

Mackay, E.J.H. 1938. *Further Excavations at Mohejo-daro*. 2 vols. Delhi: Government of India.

Mallory, J.P. 1989. *In Search of Indo-Europeans*. London: Thames and Hudson.

Marshall, J. 1931. *Mohenjo-daro and the Indus Civilization*. 3 vols. London: Arthur Probsthain.

Meadow, Richard H. 1984. Notes on the Faunal Remains from Mehrgarh with Focus on Cattle (*Bos*). In Bridget Allchin (ed.) *South Asian Archaeology 1981*, pp. 34-40. Cambridge: Cambridge University Press.

Muller, F. Max. 1890, reprint 1979. *Physical Religion*. New Delhi: Asian Educational Service.

Nandargikar, G.R. (Ed.). 1971. *The Raghuvaṁśa of Kalidasa*. Fourth Edition. Delhi: Motilal Banarsidass.

Nichols, Johanna. 1997a. The Epicentre of the Indo-European Linguistic Spread. In Roger Blench and Matthew Spriggs (eds.) *Archaeology and Language* I: 122-48. London: Routledge.

——. 1997b. The Eurasian Spread Zone and the Indo-European Dispersal. In Roger Blench and Matthew Spriggs (eds.) *Archaeology and Language*, II. London: Routledge.

Parker, R.N. 1918. *A Forest Flora for the Punjab with Hazara and Delhi*. Lahore. (Dehradun: Bishen Singh Mahendra Pal Singh, Reprint 1973.)

Pettinato, G. 1972. Il commercio con l'estero della Mesopotamia meridionale nel III millennio a.c. alla luce delle fonti letterarie e lessicali sumeriche. *Mesopotamia* VII: 43-166.

Possehl, Gregory L. 1996. *Indus Age: The Writing System*. New Delhi: Oxford & IBH Publishing Co. Pvt. Ltd.

Raikes, Robert. 1968. Kalibangan: Death from Natural Causes. *Antiquity*, XLII: 286-291.

Ramesh Rao, K. and S.K. Purkayastha 1972. *Indian Woods*, Vol. III. Dehradun: Forest Research Institute.

Rao, S.R. 1979. *Lothal: A Harappan Port Town (1955-62)*, Vol. I. New Delhi: Archaeological Survey of India.

——. 1985. *Lothal: A Harappan Port Town (1955-62)*, Vol. II. New Delhi: Archaeological Survey of India.

Ratnagar, Shereen. 1981. *Encounters: The Westerly Trade of the Harappan Civilization*. Delhi: Oxford University Press.

Renfrew, C. 1988. *Archaeology and Language*. New York: Cambridge University Press.

——. 1999. Time Depth, Convergence Theory and Innovation in Proto-Indo-European: 'Old Europe' as a PIE Linguistic Area. *Journal of Indo-European Studies*, 27 (3-4): 258-293.

Sabeti, H. 1966. *Trees and Shrubs of Iran.* Tehran (In Persian)

Saraswat, K.S. and Anil K. Pokharia. 2003. Palaeoethnobotanical Investigations at Early Harappan Kunal. *Prāgdhāra* (Journal U.P. State Archaeology Department), No. 13, pp. 105-39.

Shaffer, Jim G. and Diane Lichtenstein. 1999. Migration, Philology and South Asian Archaeology. In Johannes Bronkhorst and Madhav M. Deshpande (eds.) *Aryan and Non-Aryan in South Asia: Evidence Interpretation and Ideology.* Cambridge: Harvard University.

Shah, Sayid Ghulam Mustafa and Asko Parpola (eds.). 1991. *Corpus of Indus Seals and Inscriptions;* 2. *Collection in Pakistan.* Helsinki: Suomalainen Tiedeakatemia.

Sharma, A.K. 1993. The Harappan Horse was Buried Under the Dune of .... . *Puratattva,* 23: 30-34.

Sharma, R.S. 1999. *Advent of the Aryans in India.* New Delhi: Manohar Publishers.

Singh, Bhagwan. 1995. *The Vedic Harappans.* New Delhi: Aditya Prakashan.

Srivastava, K.M. 1991. *Madinat Hammad: Burial Mounds 1984-85.* Bahrain National Museum.

Stewart, R.R. 1972. An Annotated Catalogue of Vascular Plants of West Pakistan and Kashmir. In E. Nasir and S.I. Ali (eds.), *Flora of West Pakistan.* Karachi.

Talbot, W.A. 1911. *Forest Flora of the Bombay Presidency and Sind,* Vol. II. Poona. (Reprint 1976, Bishen Singh Mahendra Pal Singh, Dehradun, and Periodical Experts, Delhi.0

Thapar, Romila. 1988-91. In *Journal Asiatic Society of Bombay,* Vol. 64-66. pp. 259-60.

Vats, M.S. 1940. *Excavations at Harappa.* 2 vols. Delhi: Government of India.

Watt, G. 1889-1893. *A Dictionary of Economic Products of India,* Vols. I-VI. London. (Dehradun: Bishen Singh Mahendra Pal Singh, 2nd Reprint, 1972.)

Wheeler, R.E.M. 1947. Harappa 1946: The Defences and Cemetery R-37. *Ancient India,* 3: 58-130.

——. 1968. *The Indus Civilization,* 3rd edn. Cambridge: University Press.

Whitaker, Thomas W. and W.P. Bernis. 1976. *Cucurbits: Cucumis, Citrullus, Cucurbita, Lagenaria (Cucurbibaceae).* In N.W. Simmands (ed.), *Evolution of Crop Plants,* pp. 64-69. Longman: London and New York.

Witzel, Michael. 1995. Ṛigvedic history: poets, chieftains and polities. In George Erdosy (ed.) *The Indo-Aryans of Ancient South Asia,* pp. 307-52. Berlin: Walter de Gruyter.

SANSKRIT TEXTS

Ṛigveda Saṁhitā, edited by Shriram Sharma Acharya and Bhagvati Devi Sharma, Brahmavarchasa, Shantikunja, Hardwar, 1994-95.

Kṛiṣṇa-Yajurvedīya Taittirīya Saṁhitā, edited by Shripad Damodar Satvalekar, Pardi, 1957.

Bhāvaprakāśa Nighaṇṭu of Śri Bhāvamiśra, edited by G.S. Pande and published by Chaukhamba Bharati Academy, Varanasi, 1995.

Mahābhārata, General editors V.S. Sukthankar and S.K. Belvalkar, Bhandarkar Oriental Research Institute, Poona, 1948.

Yajurvedīya Maitrāyaṇī Saṁhitā, edited by Shripad Damodar Satvalekar, published by Svadhyaya Mandala, Aundh (Distt. Satara), 1942.

Raghuvaṁśa of Kalidasa, edited by Gopal Raghunath Nandargikar and published by Motilal Banarsidass, Delhi, 1971.

Śatapatha Brāhmaṇa, Hindi translation by Ganga Prasad Upadhyaya, edited by Swami Satyaprakash Sarasvati and published by Govindaram Hasananand, Delhi, 1998.

Śivamahāpurāṇa, Sanskrit text by Śrikṛiṣṇa Dvaipāyana, translated in Hindi by Jwala Prasad Misra and published by Khemraj Shrikrishna Das, Bombay, 1988.

INDEX

SANSKRIT INDEX

ENGLISH INDEX